I0759777

Sails and Shadows

The publisher and the University of California Press Foundation gratefully acknowledge the generous support of the Constance and William Withey Endowment Fund in History and Music.

Sails and Shadows

How the Portuguese Opened the Atlantic and Launched the Slave Trade

Patricia Seed

UNIVERSITY OF CALIFORNIA PRESS

University of California Press
Oakland, California

Library of Congress Cataloging-in-Publication Data

Names: Seed, Patricia, author.
Title: Sails and shadows : how the Portuguese opened the Atlantic and launched the slave trade / Patricia Seed.
Description: Oakland, California : University of California Press, [2026] | Includes bibliographical references and index.
Identifiers: LCCN 2025015773 (print) | LCCN 2025015774 (ebook) | ISBN 9780520415874 (cloth) | ISBN 9780520415881 (ebook)
Subjects: LCSH: Slave trade—Africa, West—History. | Navigation—Technological innovations—Portugal—History. | Navigation—Atlantic Ocean—History. | Seafaring life—Portugal—History. | Portugal—History—Period of discoveries, 1385–1580.
Classification: LCC HT1332 .S44 2026 (print) | LCC HT1332 (ebook) | DDC 381/.44089691009024—dc23/eng/20250616
LC record available at https://lccn.loc.gov/2025015773
LC ebook record available at https://lccn.loc.gov/2025015774

Manufactured in the United States of America

GPSR Authorized Representative: Easy Access System Europe, Mustamäe tee 50, 10621 Tallinn, Estonia, gpsr.requests@easproject.com

34 33 32 31 30 29 28 27 26 25
10 9 8 7 6 5 4 3 2 1

Contents

Illustrations

MAPS

FIGURE

Introduction

By 1400, sailors around the world had conquered all oceans except one. Ancient Greeks had sailed the inland waters of the Mediterranean and Black Seas. Persian, Roman, and Arab navigators had followed the east coast of Africa about two-thirds of the way south and trailed seasonal winds to India. Asian ships had crossed the South China Sea and voyaged around the Bay of Bengal to Arab settlements along the East African coast. Intrepid Polynesians had traversed thousands of miles of Pacific waters both east and west, occasionally north and south. But despite the immense reaches of the Atlantic Ocean, only the Norsemen had crossed a tiny sliver in the far north.[1]

Thus, no one outside the Americas knew of the Western Hemisphere—because, before the Portuguese voyages, nobody could cross the Atlantic Ocean. "One cannot sail the [Atlantic] Ocean," declared a Roman writer. "Beyond the Ocean, nothing." Another Roman concluded that if "any habitable lands [rested] either in the Ocean or beyond the Ocean . . . it was impossible to reach them."[2] Nearly a thousand years later, a Muslim writer expressed almost identical sentiments: "No ship sails therein, nor is any habitable land there."[3]

Likewise, the Iberian writer al-Masudi expressed the dread that prevailed among Roman and Muslim writers: "Observe . . . the warring of its winds," he wrote. "Such [is] the mad rage of the sea when stirred from its lowest depths: there is no anchorage at hand, no safety: all is unknown." A Roman orator proclaimed: "The waves are enormous. . . .

The winds are violent . . . on all sides, savage sharks and dogs of the sea, seiz[ed] their ships . . . [leaving the sailors] stranded . . . to be mangled by the wild beasts of the sea."[4] Even modern geophysicists agree that "the Atlantic is arguably the ocean's most complex basin."[5]

Portuguese sailors were the first to overcome the obstacles to crossing the world's second largest ocean. Still, only two of their successful Atlantic journeys—those of Bartolomeu Dias and Vasco da Gama—have remained well known.[6] The first proved that a waterway linked the Atlantic and Indian Oceans; the second created a novel sea route from Lisbon to centuries-old trade in a small corner of the Indian Ocean. When recounted so matter-of-factly, these two events appear to have come out of nowhere. So, how did these two expeditions navigate nearly four thousand miles south and almost three thousand miles west when no one else in human history had advanced south of Newfoundland?

Sailors' fears aside, the Atlantic presented uniquely challenging winds, waves, and weather. To sail south beyond the Canary Islands had proved impossible for millennia because broad currents flowed west, not south. Major currents near the equator reliably traveled east or west in other oceans. Yet sections of these major currents in the Atlantic reversed for a couple of months, only to return to their previous path. Furthermore, nearly all existing sailing techniques, except those of the Polynesians, worked only in the Northern Hemisphere.[7] None would allow ships to travel north and south of the equator. In short, navigating the Atlantic required significant changes to sailing.

However, while embarking on these breakthrough journeys, the Portuguese also inaugurated new relationships. When sailors needed to refill their drinking-water barrels, acquire food, or make temporary repairs, they often anchored near familiar people along the coast. However, they reacted differently than other Europeans. Rather than seizing the opportunity for a significant land grab like the English or controlling people to use their labor like the Spanish, they sought to extract human beings, spices, and gold from these communities and exploit their maritime prowess to bring these goods and people back to Portugal to be sold at a profit. Although few records of these initial interactions survived, they illuminate the darker side of these epoch-making voyages.

To learn how Portuguese sailors acquired the knowledge that enabled da Gama and Dias to travel such lengths across the Atlantic, most people would look for written reports or maps of earlier voyages to understand how Portuguese navigators mastered the uniquely challenging Atlantic conditions. However, while such research has yielded a handful of documents, no trove of firsthand accounts has been found explaining how they approached sailing the ocean that no one had crossed.

Unfortunately for historians trying to understand these developments, an unprecedented natural disaster in 1755 wiped out a wealth of documentary evidence detailing how pilots overcame the half-dozen major obstacles to sailing the Atlantic. Without the usual written sources of information, it has been impossible to authenticate the reach of Portuguese sailing in the Atlantic. Thus, this account begins with the cataclysmic events in Lisbon on November 1, 1755: an earthquake and tsunami, the implications of which would radiate across Europe.

1

Nature Intervenes

On a seemingly ordinary morning, the first of November 1755, the city of Lisbon was thrust into chaos by twin disasters. Shortly after 9:30 a.m., the ground beneath quivered for a minute and a half, then abruptly ceased. Moments later, it trembled again in a startling turn of events, this time with greater intensity. Pausing momentarily, the earth then convulsed violently for the next three minutes, obliterating the city's many multilevel masonry buildings and trapping or killing those within. Grand cathedrals' majestic, vaulted stone ceilings crumbled, crushing the faithful at All Saints Day Mass. Terrified Lisbonites sprawled on the ground as the earth shook beneath their feet. Recovering from the shock, many attempted to escape through the city's narrow streets. However, the escape route was fraught with danger. Large chunks of stone broke free and tumbled down from homes, the city's massive warehouses, and the grand palaces, blocking entire streets. Shaken survivors struggled for breath as the crumbling stonework filled the air with fine white dust, obscuring their vision. Yet a second calamity was about to unfold, equally devastating and even more unexpected.[1]

At around 10:45 a.m., just over an hour after the earthquakes had begun, those gathered in the plaza saw something happen that neither they nor any living European had ever seen. At first, the mile-and-a-half-wide waters of the Tagus started to recede as if being pulled back toward the Atlantic by an invisible hand. Observers on the higher ground then watched in horror and disbelief as a swirling mass of foaming brown water roared back up the river and into the harbor.[2]

As a nearly hundred-foot-high column of salt water entered the Tagus River, it crashed through the buildings on the river's edge at around thirty miles an hour. It had quickly crushed both the royal mint and the shipyard, filled with oceangoing boats under construction. Next it demolished the king's multimillion-volume library (the Riverside Palace), buried the commercial buildings along the central plaza, and flooded all the warehouses holding goods from overseas trade. All within minutes.

While official government documents were usually stored safely elsewhere, practical sailing guides and charts were kept near the wharves, within easy reach of arriving and departing ships.[3] Given that only two or three ships were dispatched simultaneously to explore unfamiliar seas, few copies of those documents were needed. Each pilot required a single record book in which to note his judgments on new routes and sketch new coastlines. In the first decade of the sixteenth century, the Crown authorized over a hundred ships to sail the Atlantic, making it necessary to create dozens of handwritten copies of navigational instructions—only two of which have survived, mislaid and discovered only near the end of the nineteenth century.[4] When printing began halfway through the fifteenth century, typographers prioritized the safe storage of religious books, with their ready audience, leaving handwritten fifteenth-century navigational records near the harbor, where they were uniquely vulnerable to the tsunami. In Portugal, the surviving fifteenth-century navigational evidence consists of two copies of mathematical calculations, an occasional sheet in government files, and outside official contacts from Congo, with only scattered, fleeting references appearing in authorized histories.[5]

Most fifteenth-century evidence has been found in other countries. A German printer had smuggled three brief accounts and one nautical guide out of Lisbon in 1507. A temporarily employed Venetian published his story in Italy that same year. A discarded history of the first fifteen years was located in a Paris library in 1840. None of the documents found elsewhere include information on the original journeys south of the Congo River, leaving only skeletal details on the most extended and daring voyages across the Atlantic and southeastern Indian Oceans during the final two decades of the fifteenth century. A hundred years after the initial exploratory journeys began, general guides appeared in print, based on improved timing and routing of ships. Still, none revealed how the original challenges were surmounted.

One hundred years after the first voyage succeeded and the economic success of the voyages was fully established, Portuguese authors first published histories of the previous hundred years to celebrate their country's past.[6] These volumes were written for audiences uninterested in either the technical advances in sailing or the darker side, the slave trade. Moreover, these accounts discuss the voyages in general terms, while frequently including contradictory information from unverifiable sources. None of these elites (as all writers were at the time) sought to study the actions and innovations of mere pilots or how their technical breakthroughs occurred.

Thus, while the sources may be lost, this book's central task is reconstructing lost navigational resources while understanding their implications for the Age of Sail. This task will take us forward five hundred years to the remarkable recent advances in climate science that have vastly improved our understanding of the behavior and interactions of oceans and weather.

Moreover, these sciences offer a wealth of data. In recent years, the rise in global temperatures has focused attention on the ocean, the principal regulator of Earth's temperature. The ocean absorbs most of the heat from the sun and cools the planet to levels compatible with human life. To understand how much the world has warmed historically, geoscientists have amassed immense quantities of data from the past, offering new opportunities to understand the seas and weather experienced by the first Portuguese ships to sail into the uncharted waters of the Atlantic.

Likewise, two twentieth- and twenty-first-century innovations have vastly improved our understanding. The first is our relatively new ability to see the oceans from space. In 1972, the United States launched the first Earth observation satellite, and today, vast arrays from many nations orbit Earth, providing detailed measurements of sea levels, air temperatures, winds, and currents. Concern for the changing climate has also increased the number of in-depth regional scientific studies. For example, a recent examination of moored buoys off the coast of Angola revealed that currents from opposite directions clashed for centuries—and thus explained why a highly skilled Portuguese pilot had to turn back repeatedly when first attempting to sail south of the Congo River. The second breakthrough is the exponential growth in the processing power of computers, which has advanced macro- and micro-level knowledge of the Atlantic and southern Indian Oceans. In 1946, the fastest computer could make only two hundred calculations per second;

by June 2020, supercomputers were 50 quadrillion times faster. Given that the world's oceans and weather interact so intricately that they demand such superfast (exo-scale) computers, we have more and more opportunities to simulate models of the sea and atmosphere. Most people learn only indirectly how effective these advances have been—when they see the increasingly accurate predictions of hurricane and typhoon paths, which are possible only through massive computing power. Such powerful tools help us understand the unusually complicated Atlantic.

Encountering these multiple maritime challenges posed by the Atlantic, Portuguese sailors developed six new nautical methods. First, they had to innovate a way to round the previously impassable "bulging" cape in the mid–North Atlantic by inventing a sailing technique, still widely used, called "current sailing." Second, as they ventured farther south, they had to understand the southern boundary of the Gulf Stream to return to Portugal. Third, they had to figure out how to estimate the timing and often shifting locations of unpredictable calms that stranded ships across a broad swath of the Atlantic. Fourth, they faced deciphering the uniquely complex crisscrossing Atlantic equatorial currents. Fifth, to sail south of the equator, they invented the first scientific technique for determining latitude anywhere on the globe. Finally, for Vasco da Gama to reach India, they had to discover the Coriolis effect centuries before its physics was understood.

The desperation of foreign governments to acquire Portuguese techniques illustrates their efficacy. European powers copied Portuguese maritime strategies through the use of spying, theft, and bribery. Italy's Duc de L'Este bribed an experienced map maker to secretly copy the master map of all newly discovered coasts. French merchants bribed low-level Portuguese sailors, Spaniards recruited disaffected Portuguese scientists who took their research across the border, and Englishmen copied the resulting Spanish texts.[7] However, these imitators only needed to know how to use their methods, not understand how they were devised. That information remained in archives along the river, lost in the mud, debris, and ashes left behind after the tsunami, earthquake, and fires.

Thus, to reconstruct that painstaking process of discovery, we must examine the ordeals and wonders Portuguese sailors faced as they found a path down the African coast.

Even when sailing far from land, ships had to periodically find harbors to resupply their fresh water and food and repair their vessels. Searching for these mandatory stopping voyages also allowed Portuguese sailors to see uninhabitable deserts, muddy swamps, vast forests, deep rivers, shallow intermittent streams, steep scrub-covered hillsides, flat sandy beaches, and needle-like rocky shores when the maritime environment permitted. Some landscapes were uninhabitable, while others created unique ecosystems seen from aboard ship for the first time.

Such environs contained novel plants, mentioned only as asides. But these random references to flowering plants or fruit-bearing trees provide unexpected clues to the months of a ship's passage when all other clues are missing. Nature also provided Indigenous peoples with plants previously unknown in Europe. Some formed the basis of healthy nutrition; others contained uniquely deadly chemicals that provided Indigenous people an advantage over armed Portuguese men. Indeed, many of these coasts in equatorial zones were inhabited, contrary to what generations of scholars had claimed.[8] Yet different ecologies shaped the needs, expectations, customs, and goods of the diverse communities these seamen encountered. Horse-riding nomads lived differently than swamp-dwelling rice growers, and oceangoing fisherfolk operated differently than coastal oyster and mussel collectors.

When direct slave-taking raids on these African communities floundered, Portuguese voyagers began to search for local partners who would undertake the challenging task of rounding up people to sell to visiting Portuguese ships. Upon encountering goods with a potential market in Europe, such as gold or spices, the Portuguese quickly bundled and loaded them on the ships. To maintain good relations with the local elites supplying such goods, Portuguese visitors allied not with the traditional social elites but with others, more willing to organize slaves and goods in exchange for European goods, creating a new nexus of power in these societies. To sustain these commercial connections, the Portuguese valued understanding the changing desires of local elites for products they could bring to trade. A unique part of that strategy stemmed from a curiosity about local communities and African languages. From the first capture of slaves, members of Portuguese expeditions volunteered to spend time residing in Indigenous communities, learning the languages and customs. Nonetheless, this almost anthropological interest in understanding newly encountered societies by learning their languages and knowing more about them than their trade rituals was overshadowed by the (simultaneous) operations of the forerunners of the brutal Atlantic slave trade.

In a culturally distinctive way, these Portuguese initiated contact with Indigenous peoples in Africa, leading to their large-scale participation in slave trading and an uncanny competence in the dynamics of cultural contact. Their scientific achievement and ethnographic curiosity resulted in the moral ambiguities of the distinctively Portuguese role in opening the Atlantic for seaborne commerce in goods and people and, ultimately, in their creation of the technologies Europeans used when they began sailing the globe. This remarkable story reveals a great deal about the complex interplay between technology, markets, and the unique colonial strategies developed by the Portuguese—strategies that would set the stage for both the Age of Sail and the era of Atlantic slavery that would endure throughout the following centuries.

2

Around the Bulge

Bojador, 1434

To truly understand the story of the earliest journey, we must examine the natural barriers that blocked European progress into the Atlantic for so long. Thus we turn to West Africa, where the morning fog lifts, revealing an unremarkable red dune off the western coast of the Sahara, just south of Morocco. Tranquil waters lap at its gently sloping base as schools of sardines surface, their silvery skins giving the illusion of waves.[1] For centuries, though, this ordinary sandbank had a terrifying reputation. Vessels fortunate enough to reverse course upon sighting the red dune survived, but those that attempted to sail around it disappeared, never to be seen again.

Yet a glance at a map or satellite picture does not explain its deadly reputation, for most would find it nearly impossible to pick out the so-called Bulging Cape (Bojador). The mystery and intrigue surrounding this cape, which appeared very flat and ordinary on maps and photographs, was part of its deadly charm. Two threats confronted anyone who would sail south, the first hidden below the surface. From the cape's base, a shallow bottom extends twenty miles westward and sixty-five miles south into the Atlantic, making it the most extensive sandbank off the eastern side of any sea or ocean in the world.[2] Any boat heading to the base of the red sands would soon become stuck in the shallows—giving the cape its other, five-hundred-year-old nickname: the Cape of Sand, the sand monster.[3]

Sailing out beyond the shallow waters, though, posed the second threat. While sailing around the cape, sailors lost sight of the red sand-

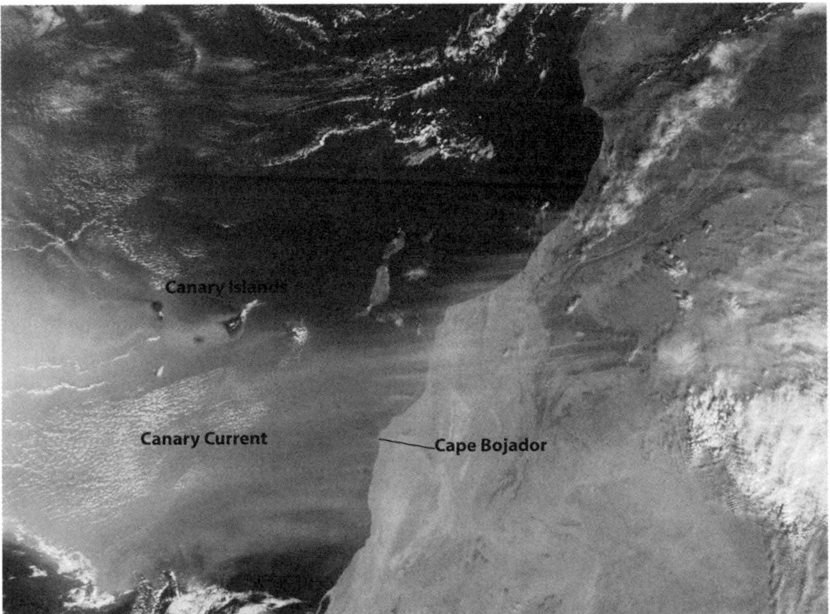

MAP 1. Cape Bojador and the Canary Islands. NASA.

bank. The current that had obligingly carried them away from the shallows then trapped them inescapably in an immense current that modern measurements have determined extends over a thousand kilometers (620 land miles) to the south. Today, it bears the name Canary Current, after the Atlantic islands near the African coast where it originates. As one fifteenth-century author noted, "from this cape forward, no one ever navigated because of the massive currents," whose grip ships of the time lacked techniques to escape.[4] Instead of allowing ships to move south, this massive, slow-moving flow headed steadily southwest, away from the African coast, bearing any boat away with it. Making the situation worse, the farther the ship moved from the coast, the slower the current traveled.

In the Northern Hemisphere, southward currents amble and take up much space. Off the Bulge, the Canary Current becomes one of the slowest-moving currents on the planet. It initially drifts southwestward at a leisurely three to seven kilometers (1.8–4.0 miles) over an entire day, while its northbound counterpart zips through the same distance in an hour and a half. As if to worsen the situation, it moves marginally more quickly closer to the impassable shallows, only to grow lethargic farther at sea. By sailing west to avoid the sandbank, an unprepared crew would become trapped by this vast, increasingly sluggish current for agonizing days.

If, by some miracle, the sailors managed to survive, more danger awaited. A short distance from the coast, the Canary Current turns west. From there, it slowly slides into the North Equatorial Current, which continues westward to the Caribbean. Victoria Murden McClure, the first woman to single-handedly row across the Atlantic, found the Canary Current so helpful that she declared that a barrel thrown overboard into the current would wind up in the Caribbean.[5] That was precisely how Christopher Columbus reached the New World in 1492: He caught the Canary Current, whose equatorial continuation carried him directly to the Caribbean. But those unprepared for a lengthy westward voyage would find themselves pulled by this slow-moving, thousand-kilometer-wide current, only to then be trapped by the North Equatorial Current, which would relentlessly haul them westward for miles, eventually exhausting any food and water as they slowly succumbed to dehydration and starvation. Even today, small vessels of migrants sometimes become trapped in the Canary Current, only to wash up in the Caribbean or Brazil with few, if any, survivors.[6] In 2021, more than a thousand people met this fate.[7]

In short, the currents made traveling one way to the west feasible, but made sailing south into the Atlantic impossible. The current that flowed around the cape's broad sandbanks remained a second insurmountable barrier for those desiring to travel south rather than toward the Caribbean. To illustrate the difficulty of conquering the Canary Current, one need only consider that nearly every other ocean had been navigated by this era. In the Western Pacific, Polynesians had succeeded in crossing thousands of miles. Africans, Persians, and Romans had traveled the small northwestern corner of the Indian Ocean and adjacent coast for centuries. Traversing the northernmost sliver of the North Atlantic and elsewhere, ships clung to the shores like crabs in inland seas such as the Baltic and Mediterranean, and along northwestern European and Southeast Asian shores. Yet none of these courageous ocean sailors had ever succeeded in navigating farther south than 650 miles beyond the straits of Gibraltar.[8]

It was not for lack of trying. For thousands of years, ships had attempted to overcome the Bulging Cape, Bojador. Phoenicians, Greeks, and Romans had failed, as had Arabs, Vikings, and other Europeans. From the wealthy medieval African kingdom of Mali, for example, a well-supplied fleet of two thousand ships set out to cross the Atlantic, never to return. Anyone who attempted the crossing—deliberately or accidentally—was carried one way by winds and currents, to the Caribbean or even the eastern coast of South America, never to return to their homeland.[9]

A geological report exposes the best-known tale of a successful journey around Bojador as a fantasy. A Phoenician navigator named Hanno was said to have rounded the cape and then sailed between islands where several volcanoes were erupting. However, only one place along the western African coast had multiple active volcanoes during the past four millennia, and those eruptions occurred circa 2000 BCE in the Canary Islands, which therefore were Hanno's most probable destination.[10] In other words, Hanno stopped short of Bojador.

After setting out, sailors met the impenetrable darkness of the Atlantic described in centuries of Roman and Muslim writing. A well-known Roman orator declared that in the Atlantic, "the light is mingled in dense gloom; . . . [for] stars there are none, or they are unknown. . . . The eye cannot pierce the darkness on all these waters."[11] Muslim writers labeled the scene with similar phrases, such as "gloomy sea" and "sea of darkness."[12] Wrote one scholar, "The waters of this sea are covered with clouds and dark in color. The waves are enormous, and the sea is deep. Darkness reigns continually, and navigation is difficult."

While such observations underscore the difficulty of the challenge, a more profound problem with early attempts was traditional navigation practices. The countless sailors who had successfully navigated elsewhere tended to follow rather than fight the currents. In some places, currents continuously circle, allowing easy passage in the Bay of Bengal and the straits between Greenland and eastern Canada. Elsewhere, flows reverse seasonally, as along the northeast coast of Africa, in the northwest Indian Ocean, and in the East and South China Seas. Contending with a broad, drawn-out, opposing current demanded an entirely new approach.

AROUND THE BULGING CAPE

The fifteenth century saw the Portuguese challenge the North Atlantic. Around 1422, Dom Henrique (fourth son of King John I), who became known as Henry the Navigator, encouraged sailors to attempt the Bulging Cape.[13] Over the next decade, many ships set out from Lisbon; some turned around before reaching the deadliest part of the currents, while others disappeared forever into the mid-Atlantic. And although the prospects may have seemed grim, the prince continued to find many willing to take their chances. Portuguese poet Fernando Pessoa characterized the anguish of those dozen years between the first deliberate Portuguese attempt to cross Bojador and eventual success with these words: "Whoever desires to go beyond Bojador / Must go beyond the pain."[14]

Then, one spring in 1434, a single small Portuguese boat returned from around the cape with seven men aboard, to general surprise and disbelief.[15] Prince Henry, who had authorized the trip, refused to believe that navigator Gil Eanes, a mere commoner, had succeeded. He demanded that the boat's captain immediately repeat the journey, accompanied by a second ship carrying a trusted observer, Antonio Baldaia, whom the prince deemed of sufficient social standing. The matter was settled when the two boats and crews returned.[16] A way to sail across the massive currents surrounding Bojador did exist. However, Eanes's rounding of Bojador in 1434 remains one of the least understood of ocean crossings. While many history books and other texts correctly report Eanes's achievement, they falter in explaining *how* he managed to sail southeast around the Bulging Cape and south along Africa's coast, where others had only met disaster. The official histories leave us to guess what his uniquely successful methods may have been. However, armed with a smattering of oceanography, we can begin to judge the timing and characteristics of Eanes's winning strategy.

The solution to sailing around Bojador proved counterintuitive. Sailors the world over followed currents. To reach the other side of the cape, a ship would have to sail east against two obstacles: a northern wind and a southwesterly current. While sailing into the wind was well known to sailors, sailing against an ocean current was significantly more complex than going against the wind. Importantly, a single knot of current is equivalent to thirty knots of wind—in land terms, movement of 45.3 miles (73 kilometers) per hour. Today, the counterintuitive approach to the crossing that Eanes (or his pilot) most likely invented is called current sailing.[17]

Current sailing accounts for the effects of current on a ship's direction. To do so, Eanes, or those working with him, would need to have invented a method of measuring the speed and direction of the current. The current moved slowly past Bojador (0.03–0.07 meters/second), as did most currents along the ocean's eastern edges. But closer to the shores of Bojador, the current picked up, making the initial southeastward turn the most challenging. Hence, the current's speed would need to be checked repeatedly.

To establish the speed and direction in which the current pulled his ship, Eanes or his pilot would have stopped to measure the current's speed and direction, or he may have employed another practice still used today: Some sailors improvise with a weighted stick or object that can be held below the ocean's surface to reach the current. By dropping such an item into the current near a fixed object, such as an anchor chain, and then timing the distance and direction that the dropped object travels, a pilot can quickly calculate the velocity and direction of the current.

Because of this continual pull southwestward and away from the African coast, the boat's direction and the actual course it traveled, or "course made good," would differ. Once the velocity and direction of the current were known, the pilot would use a triangle to estimate the angle at which the boat needed to sail on the first leg of the trip to its destination. When heading east toward the coast, Eanes or his pilot would have needed to steer his boat northeast in the direction to compensate for the current. After sailing a certain distance, the ship had to compute the difference between the estimated and actual current that acted on the boat. If the vessel had traveled too far south, the course would need to be corrected to sail more to the northeast. The pilot would also measure the speed and direction of the current at this second location and include that figure in calculating the direction for the next section of the trip. This process would be repeated until the destination was reached. By acquiring this information, Eanes could have continually corrected his course to compensate for the southwestward drift and changing current speed. Unlike previous guides to sailing that used prefixed numbers, current sailing requires mathematical literacy and the ability to recompute directions.[18]

Thus, Gil Eanes became the first successful current sailor, his innovation depending on careful measurement of the direction and speed of ocean currents. As other Portuguese sailors began heading south, gradually conquering the previously unexplored mid-Atlantic and South Atlantic, they paid great attention to measuring both currents and winds—employing an innovative strategy and thereby succeeding where none had previously. In short, Eanes succeeded by making continuing calculations.

His use of mathematics to conquer Bojador was surely one of the first cases in which a pilot used mathematical techniques while sailing. Of course, pilots had devised techniques and implements to help them navigate everywhere else. Arabs and Persians had studied the apparent movement of the stars to navigate the Indian Ocean, and Micronesians and Polynesians had memorized the names and paths of hundreds of stars. Arab sailors had adapted a Chinese directional device for moving across the Mediterranean, and this instrument, the compass, soon appeared on ships throughout the region. However, in the historical record, none of these navigators had made fresh mathematical calculations during their voyages. Rough estimates and measures based on the varied sizes of pilots' hands were accurate enough to make sailing possible in these oceans. Even the Arabs, whose mathematical skills were legendary, never applied that knowledge to the craft of navigation. The most likely reason, of course, was that they did not need to. Approximation techniques sufficed for vessels traveling between East African and Indian ports, making mathematics unnecessary. Yet mathematics—including the use of triangles and other geometry (and, later, trigonometry and even physics)—would become crucial to nearly all subsequent Portuguese breakthrough crossings of the Atlantic, including the storied voyages around the Cape of Good Hope and the battle against the world's second fastest current, the Agulhas, on the trip north to the well-traveled northwestern quadrant of the Indian Ocean.[19]

For historical events, the significance of a particular moment becomes apparent only in retrospect. Although applauded at the time, this breakthrough voyage rounding the Bulge became a landmark achievement only a decade later. Eanes's ship returned with no captured people, only a handful of flowering plants pulled up from nearby shores.[20] The only flower in the area that produces purple blossoms is *Fagonia cretica*, called the Rosa de Santa Maria in Portuguese, which blooms from December to June. These flowers were a harbinger of the wildly unfamiliar world that the Portuguese would reach and to which they would subject ruinous violence.

3

From Mistaken Expectations to Conquests

The Sahara, 1434–1444

As the magnitude of Gil Eanes's discoveries became apparent in the following decade, Prince Henry and his courtiers finally grasped their profound importance. At first, unaware that the southern expeditions he sponsored were reaching untested waters, Henry viewed them only as an opportunity to find local sailors already operating south of the Bulge. However, when the ships initially failed, repeatedly, to locate sources of informants—eventually finding only small groups of desert nomads and coastal fishermen—Henry invoked a military method he had previously exercised in North Africa: the Iberian tradition of "conquest." This would prove a pivotal moment, when the Portuguese cited this historically legitimized Iberian basis for warfare—which had justified the kidnapping and enslavement of Muslim adversaries—in order to begin legally enslaving any African people Henry's authorized ships encountered.

Although many Portuguese writers praised the prince's supposed piety during this period, a close reading of even his chosen chronicler's account of his behavior during this phase starkly exposes the prince's superficial devotion and fundamental lack of empathy. Shortly after embracing the conquest strategy, he broadened the targets of conquest to include non-Christians and soon thereafter abolished long-standing traditions that allowed the ransoming of captives. Initially facing extensive political opposition to enslaving simple non-Muslim fishermen, Henry had to repeatedly enlist the assistance of Pope Eugene IV to justify enslaving people. These drastic changes made Prince Henry the inventor of the trade in enslaved African persons as chattel, a dark chapter of human history played out against the stark Saharan sands.

After four uneventful early voyages, the nautical narrative also changed. As they ventured farther south, Portuguese ships reported increasingly daunting challenges upon returning to Lisbon. These unexpected and arduous experiences led to a temporary cessation of the southward voyages. Before they could resume, the pilots had to master two new skills: navigating winds around a high-pressure system and locating an opposing current in the North Atlantic. With these new tools in their arsenal, the voyages resumed in 1441.

Unfortunately, historians often evoke the language of conquest in framing their narratives of Portuguese voyages along the African coast. This chapter will critique this use of *conquest*, a word introduced only gradually into the language of voyaging, and demonstrate its profound connection to the origins of the slave trade.

MISTAKEN INITIAL EXPECTATIONS

After the second, socially connected captain had confirmed Eanes's remarkable feat, Prince Henry accepted that the seemingly impassable Bojador had been crossed, paving the way for expeditions farther south. However, the initial explorations along the coast were met with disappointment. The pale sands of the Sahara extended as far as the eye could see and seemed to be devoid of any human habitation.

Henry's initial expectations of contact were high. He initially requested information about the land's geography, flora and fauna, existing sea routes, signs of habitation, and resources. Assuming that the well-known Arab camel caravans also transported their goods along the ocean, he declared that it "was certain that there were people here [south of the Bulge] who traded using camels . . . and the baggage they carried *had* to travel by sea." He expected to find scores of Arab merchant ships sailing along the West African coast, their northward journeys hampered by the Bulging Cape. These expectations, however, were far from the reality of the African coast.

Furthermore, Henry believed that, at a minimum, fishing communities would have settled along the coast, given the fact that well-established cities with extensive fishing activity—including Agadir, Essaouira, and Safi—lay north of the Bulge. Thus, likewise assuming that his ships would find the same sheltered harbors and availability of fish south of Bojador, the prince wrote that the inhabitants would certainly possess some knowledge of the sea to sustain the Portuguese seamen. Therefore, he initially asked his skippers to quiz residents on how to navigate the waters south of the Bulge.[1] However, the coast's geography doomed Henry's quest to seek out sailors and fishermen. Although its smooth coast began north of the Bulging Cape, the Saharan shoreline extended 350 nautical miles beyond the Bulge, with no harbors to shelter fishing vessels.[2]

DISEMBARKING ON HORSES AND SEIZING INDIGENOUS POSSESSIONS

Within two years of the first voyage to round Bojador, Prince Henry began to endorse a limited use of force, ordering a new expedition by Antonio Baldaia "to take a prisoner whether a man or a woman or child to satisfy the will of his master [Henry]."[3] Given that no humans had been sighted along the coast in previous travels, Baldaia sailed 136 miles beyond Bojador, bringing a pair of horses to ease travel across the sand. After anchoring, he sent two seventeen-year-olds, Heitor Homen and Diogo Afonso Almeida, to search inland on horseback for human life. When they returned to the ship, the teens reported that they had fought with a small encampment of nineteen men until nightfall. Baldaia accompanied the two young men back to the campsite the following day but found it deserted; they grabbed everything left behind to take back to Lisbon as evidence of human occupation. This first known landing of horses in the Sahara south of Bojador led to its name, the Anchorage of the Horse(s), a label it often bears today.

After seizing the goods abandoned at the campsite, Baldaia continued another 130 miles southwest, where he spied a giant stone, which he named Dorsal Fin Rock (Pedra da Galha) for its resemblance to the upper fin of a shark. However, the shape is now said to "look like a vessel under sail" rather than a shark fin.[4] On the beach opposite, he spotted another sign of human occupation along the shore. Drying on the beach near a distinctive large rock were some remarkable fishing nets, which he seized. Baldaia had thus obtained evidence that people resided near the shore south of Bojador, yet he had failed to seize a single person and carry out Henry's mission. In 1437 and 1438, the prince sent two more ships to explore the coast, but the first was forced to return by lousy weather and the second found only a monk seal colony. Baldaia had seen the latter 175 miles south of Bojador at the base of the narrow Rio Douro peninsula, where the crew collected seal skins.[5]

However, circumstances delayed further exploration of the lands south of Bojador.

AFTER THE MISSING YEARS, DISCOVERY OF A RETURN ROUTE

A four-year gap appears in the reported expeditions following Baldaia's failed kidnapping. While most historians have filled this gap with the only texts that survive, describing a political battle over the regency, there is a more straightforward explanation: the increasing difficulty of returning from the tip of the Rio Douro peninsula in the Sahara. The earliest return journeys had only required reversing Eanes's current strategy. However, farther down the shoreline, voyagers met challenging weather conditions. The Portuguese did not report the details of failed voyages at this time, so a long series of lost ships most likely explains the gap. The record picks up with a report on a new nautical course called the "Long Return" (*volta ao largo*), an extended return route that, in a counterintuitive strategy, set off to the northwest, deeper into the Atlantic. This pathway used another recent nautical discovery, one rarely connected to voyages down the African coast.

In 1432, two years before Eanes rounded the Bulge, Portuguese sailors managed to navigate back and forth between Lisbon and an uninhabited Atlantic volcanic island 755 nautical miles west.[6] They reached the island by using the top of the westward-heading Canary Current, which flowed past its southern side. However, just to the island's north lay a very different current (part of the Gulf Stream) that allowed them to return directly to Portugal.

North Atlantic currents flow in a giant clockwise circle. The best known of these is the Gulf Stream, which begins in the Caribbean and runs northward in parallel to the east coast of the United States, as far as Canada, and across the North Atlantic. However, one branch of the Gulf Stream begins to split off near North Carolina; this section streams eastward toward the newly discovered volcanic island, which was found to form part of a chain later named the Azores. Hence, this eastward branch of the Gulf Stream became known as the Azores Current.[7] Once reaching the Azores, the current weakens but moves northeastward toward the coast of Portugal.

However, the trickiest part of this Long Return was reaching the Azores from the shores of the Sahara, which meant traveling north and west, away from Africa and into the Atlantic. From the tip of the Rio Douro peninsula to that island in the Azores was 1,112 nautical miles, a

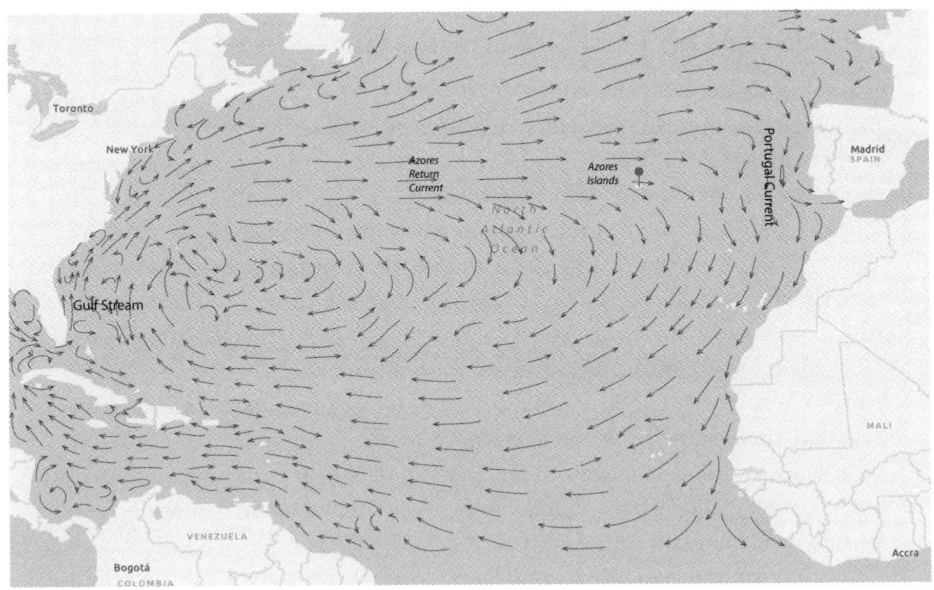

MAP 2. The Gulf Stream, the Azores Return Current, and the Portugal Current. © Esri, Tomtom, Garmin, Open Street Map Contributors and GIS User Community.

very extended return trip. However, a unique Atlantic weather pattern made the journey easier. Over the Azores lies a semipermanent atmospheric high-pressure center, known as the Azores High. Since winds circulate clockwise around these systems in the Northern Hemisphere, the zone provided favorable easterly or southeasterly winds during winter and spring, occasionally requiring tacking back and forth into the wind when the Azores High moved westward during summer and fall.

Four years later, after resolving the navigational problems of a return voyage, Prince Henry licensed another two ships. These expeditions would mark the first large-scale Portuguese capture of human beings along the West African coast and an escalation to a military stance fully endorsed by the prince. He authorized captains of the next pair of expeditions "to forcibly seize people by whatever means [they] could," an order these officers duly carried out. While none of these captures could have taken place without the discovery of the Long Return, the new route enabled Portuguese ships to reach large numbers of people, intensifying the volume of the slave trade and remaking that tradition within a decade, in what would prove a significant historical transformation.

ANCHORAGE OF THE KNIGHT: THE FIRST CAPTURES

Antão Gonçalves captained the first of two ships sent out in 1441. The prince gave him the same orders as the earlier expedition and instructed him to collect additional seal pelts. Henry then ordered a second expedition under another captain to investigate the coast south of Dorsal Fin Rock. Chafing at his limited mission, the ambitious Gonçalves set out, against orders, south of Fin Rock under the continued, mistaken belief that they would find tracks of caravans near the shore. With its explicitly exploratory mission, Henry's second ship soon caught up with Gonçalves. The combined crews went on a brief excursion into the interior, where they encountered a camp of nomads who successfully repelled their attack. While retreating to the ship, they seized a lone man leading his camel and a black woman servant, likely sent out in search of supplies for the camp. Gonçalves immediately demanded that he be given the rank of a knight.[8]

The commander of the second ship, Nuño Tristão, while understandably reluctant to grant Gonçalves the recognition he craved, agreed to accompany him on a second expedition to capture additional people. After a search into the desert, they encountered a second peaceful encampment. There, they shouted the traditional Iberian war cry, startling the residents, who fled in all directions. Several people in this previously quiet camp began to respond with their lances, but the better-equipped Portuguese killed four and captured ten men, women, and children.[9] Instead of merely approaching the groups to ask for information, armed men shouted war cries and launched themselves at these encampments, killing and wounding many. Prince Henry's official chronicler, Gomes Eanes de Zurara (in office 1454–74), who had designated earlier actions as "discoveries," labeled these new military assaults acts of "conquest."[10]

Tristão then faced a campaign waged by members of Gonçalves's crew to pressure him into granting the latter a knighthood. To keep peace and avoid risking outright conflict in unfamiliar territory, Tristão gave the ambitious Gonçalves the title. This unappealing event also left its mark on maps for decades, where the name became "Anchorage of the Knight" (Angra do Cavalheiro), marking the first time a European had been knighted while in Africa. It seems impossible to see that name today except as an expletive.[11] In any case, before returning to Portugal, the sailors wanted to ask their captives about where they resided, the types of goods they possessed, and whether they understood the sea route along the coast. This final question indicates that eight years after rounding Bojador, Portuguese voyagers still labored under the misconception that inhabitants of the Sahara transported goods by ship and, therefore, possessed knowledge of its offshore waters. However, at least one major obstacle, the language barrier, blocked them in attaining good information.[12]

ATTEMPTING TO SPEAK

None of the expedition's leaders spoke any language other than Portuguese.[13] While anchored off North Africa, where most people spoke Arabic, they turned first to the most fluent speaker on either ship, a servant of Prince Henry on Tristão's vessel, and asked him to find out all he could. After his efforts to understand one group of locals failed, three crewmen who spoke Arabic tried, again unsuccessfully. Deciding that the local language deviated significantly from the Arabic spoken in Portugal, the captains turned to a second, larger group that was present, called the Azanaghi. When Gonçalves and Tristão heard a leader of this group, named Adahu, responding in a language that sounded like Arabic, they promptly recalled the prince's servant and, having ascertained that the Indigenous elder spoke Arabic, asked both to interview the other captives. Adahu would express the captives' answers in Arabic, which the prince's servant would then translate into Portuguese.[14] These double translations yielded only disappointing information: None of these people knew anything about sea routes, and none possessed significant wealth.

THE CONQUEST STARTS: SEIZURE BY FORCE

In his official instructions of 1441, Prince Henry had authorized the captains "to forcibly seize people by whatever means" they had at their disposal. Although he may not have explicitly referred to such actions as acts of war, his captains understood them as such. The official account by Zurara characterized the man taken while leading his camel as "the first [person] made prisoner in this conquest," invoking the word *conquest* for the first time in the accounts of voyages beyond the Bulge.[15] While the word is also in the title of Zurara's book, it first appears in the text during his description of Gonçalves's and Tristão's acts of capture, thereby signaling the introduction of a military rationale.[16]

In English and European languages derived from Latin, *conquest* principally refers to combat. However, aspects of conquest in Iberia and

North Africa differed from practices elsewhere. One particularly distinctive custom involved taking prisoners, often leaders, for ransom. Spanish soldiers employed this tactic, widely accepted throughout Iberia, to significant effect in the Americas, where invaders seized the leaders of the Aztec and Inca empires—Cuauhtémoc and Atahualpa, respectively—and held them for ransom. Hernán Cortés captured Cuauhtémoc and tortured him, seeking the location of gold he believed to be hidden. At the same time, Francisco Pizarro held Atahualpa for ransom until he had filled one room with gold and another with silver. Hence, capturing individuals from the opposing side, even emperors, and holding them for ransom was integral to how Iberians conducted warfare, and they would continue employing this strategy in their overseas conquests.

A similar, unique practice formed part of their understanding of the word *conquest*. If captured individuals lacked funds or valuable objects to exchange for themselves, they could be sold as enslaved people. These practices followed a familiar pattern in what were labeled for centuries "wars of conquest" against Muslims in Spain, Portugal, and North Africa.[17] According to long-standing practices in Muslim-Christian combat, individuals captured in battle could either be held for ransom or enslaved if they lacked the means to pay. Both sides observed this custom well into the sixteenth century. One of North African Muslims' most notable wartime captives was Miguel de Cervantes, the legendary author of *Don Quixote*, who spent five years enslaved in Algeria before his family sold all their possessions and, with the aid of a charitable donation from a Spanish convent, raised the necessary funds to secure his release.[18] Both sides viewed taking their religious enemies as prisoners of war and keeping them for ransom or enslaving them as wholly legitimate.

Although Gonçalves and Tristão had launched unjustified attacks on inhabitants of the Sahara, both Muslims and Christians had long been guilty of similar unprovoked attacks, simply raiding each other's territories for either ransoms or enslaved people. Cervantes's capture had resulted from exactly such an unprovoked raid.[19]

THE FIRST REDEMPTION

To free a person in exchange for goods or money, as the Cervantes family ultimately did, constituted a historical practice called "redemption" of prisoners of war, whereby captured individuals could liberate themselves by providing money or gifts. Therefore, questioning captives about the merchandise they possessed followed the pattern of a traditional interrogation to learn whether prisoners owned valuable resources, such as gold, that they might provide in exchange for their freedom. This long-established feature of Iberian warfare was called *resgate* in Portuguese and *fidya* in Arabic, both meaning "ransom" or "redemption."[20]

After the disappointing interviews of their first captives, the Portuguese captains sent the same Arabic-speaking servant to communicate with those on shore who had yet to be captured, to find out if the community was willing to redeem any captives in exchange for merchandise or other captives. Instead of responding to this demand for ransom, community members returned to set up an ambush for any Portuguese stepping ashore. Realizing the trap, the Portuguese returned to their ships and abandoned the Arabic-speaking servant, leaving him in the hands of the local people.

This practice of kidnapping defenseless individuals unless, or until, they could be redeemed for a significant sum reveals that both Christians and Muslims possessed financial motives for taking captives. In the case of Cervantes, he was abducted from an unarmed vessel, then both sold and resold for profit as an enslaved person, and then finally sold back to his own family (for an additional profit). Hence, while supposedly acting in pursuit of legitimate religious war, individuals could and did often use this tradition to pursue financial gain. But in this case, Prince Henry's captains used the profit motive to argue for continued exploration. Gonçalves explicitly stated that this initial kidnapping constituted a "rich prize that would allow the prince to start to recuperate part of such large expenditures" in outfitting these voyages.[21] Thus, in addition to the individuals who kidnapped and held or transported the captives, another, more critical recipient of revenues from these operations was Henry himself.

Following this stalemate, Gonçalves returned to Lisbon with some of the captives. On this northward journey, the elite Adahu from the Azanaghi group pestered Gonçalves to return him to the Rio Douro region in exchange for five or six black Muslims, adding that two captured younger men held similar status.[22] Gonçalves took the matter to Prince Henry, whereupon Adahu offered to trade ten enslaved people of his own in exchange for his freedom and that of his companions. He explained that he could turn over black residents, whom he characterized as their traditional servants. Eager to return to the region, Gonçalves convinced the prince that if he paid for a return trip, it would provide an opportunity for Gonçalves to learn more in the service of royal objectives. Henry agreed and gifted Adahu with elaborate garments, hoping this would encourage an exchange of merchandise. Gonçalves then set out with his crew, the three Muslims, a German knight named Balthasar, and the king's official ambassador, who would officiate all significant "redemptions."

Both Christians and Muslims had official envoys as members of their courts, usually fluent in both languages and familiar with the customs of exchange. These officials had the same title in Arabic and Portuguese: *al-fakkāk* or *alfaqueque*.[23] Henry's alfaqueque, Martim Fernandez, spoke a variant of Arabic that all the inhabitants of West Africa easily understood. In contrast, the humble Portuguese Arab servant who had attempted to communicate with the others spoke a dialect that only one person could comprehend, suggesting that the Arabic spoken by Portuguese at court had already begun to diverge from the North African dialect.[24]

Gonçalves waited a week at anchor in Dakhla Bay, on the Rio Douro peninsula, before a single man arrived on a camel to announce that Adahu's ransom would be coming the next day. As promised, the ten male and female captives arrived the following day. Prince Henry had guessed correctly that the gifted garments would provoke a reciprocal exchange. Gonçalves received a small quantity of gold dust, a shield, and many fresh eggs, three of which made their way to Henry's table, where they were pronounced delicious. However, the only news of any import he obtained concerned a report of gold traders in the region's interior.[25]

Meanwhile, Tristão continued south according to his original mission to "travel as far beyond Fin Rock as possible." He sailed another 120 miles, as far as the White Cape (Cabo Blanco), where he found only traces of human occupation, including fishing nets like those Baldaia had retrieved several years earlier. After this disappointment, Tristão returned to Lisbon with the remaining captives—who vanished from the narrative at this point.[26] Thus, seven years after first rounding the Great Bulge and returning, Portuguese ships had managed to sail only four hundred miles, traveling generally southwest. Over the next three years, returning via the long route, they continued southwest another four hundred miles along the coastline, toward the southernmost point of the Bay of Arguin, just south of the White Cape.[27]

Off these shores, cold water surfaced, mixing with the warmer Canary Current, nourishing a wide variety of marine life. Unlike areas farther south, where nutrient-rich waters emerge only seasonally, here they surface year-round, attracting millions of fish nearly every day of the year.[28] This dependable supply of nutrition drew 80 percent of the fish-eating birds from Greenland, Europe, and Siberia to spend winter and reproduce in the area's tidal flats, as they do to this day.[29] Therefore, the desert coasts from Rio Douro to the White Cape through the Bay of Arguin attracted scores of Berber-speaking fishermen who had long congregated there.[30]

The newly arrived Portuguese viewed these fishing communities through the disparaging lens historically adopted by nearby Arab populations, who belittled this group for their diet: "They have no bread, oil, firewood, salt, or anything else that pertains to human use." While Arab groups rationalized their treatment of these poorer and darker fishers for their supposedly subhuman habits, Portuguese chronicler Zurara found a parallel with spiteful antisemitism: "These Zenagas are of an ancestry so vile and shameful that all the Moors [Muslims] consider them as we do the Jews."[31] Beginning at Rio Douro (just north of the White Cape) and continuing south, Arab groups often kept these communities in hereditary slavery that remained officially legal until 1981; in the fifteenth century, they eagerly traded members of this slave caste to the Portuguese for goods.[32]

However, would-be Portuguese slavers failed to replicate the exchange with Adahu when returning to the Rio Douro peninsula north of the White Cape. In May 1444, the Crown handed a single captain more than 2,686 meters (nearly 3,000 yards) of cloth to acquire enslaved people in Adahu's territory.[33] That same year, the Crown authorized another three ships—two of Prince Henry's and one from the king's regents—to explore the possibilities of acquiring enslaved people elsewhere near Rio Douro. Failing to make a single exchange, the three caravels returned to Portugal, carrying only an elderly Muslim man who wished to visit Portugal.[34]

However, as already noted, just south of the White Cape lay a broad protected bay. As attempts to carry out a traditional exchange with Arabs faltered along Rio Douro, the Portuguese succeeded with a more brutal strategy in the vast, protected waters south of the White Cape. In 1441, Tristão returned to the sands of the White Cape, where he had found only fishing nets; from there, he sailed south into the giant Bay of Arguin. As he approached, he saw twenty-five canoes, three or four men in each, moving their legs like oars to propel them around the shallow waters. With his crew of seven, he managed to capture thirteen fishermen, overcrowding the small caravel. As a result, some of the captives died while others fled, leaving only two. Approaching one of the smaller islands in the bay, he managed to fit these two individuals into his small boat (perhaps they were women or children, who took up less space than the men). He returned to Portugal with seventeen enslaved people

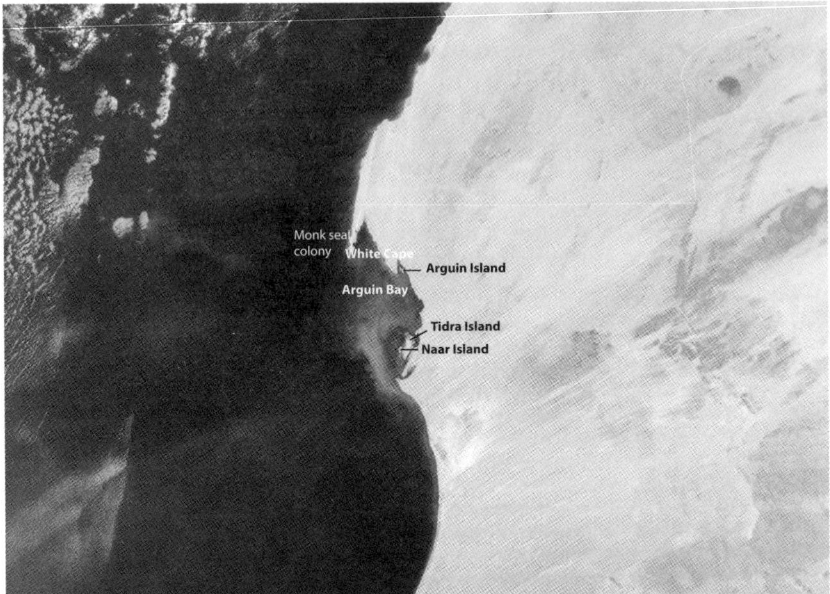

MAP 3. The Bay of Arguin. NASA.

in total, complaining that he could not take more on board because of the smallness of the ships.[35]

Although Prince Henry had sanctioned these kidnappings, this new tactic—directly seizing innocent people—would require a different rationale than the previous captures. Only one man's words could override all questions about the ethics of these kidnappings, and that individual resided in Rome.

Unlike today's popes, their fifteenth-century predecessors required the assistance of kings, emperors, and princes to maintain their papacy. Thus, they were deeply indebted to their royal political patrons and protectors and thereby amenable to ratifying requests from their loyal supporters, especially those accompanied by funds. Prince Henry submitted payment along with the language for decrees legitimizing these new seizures and, in short order, Pope Eugene IV granted Henry's initial petition.[36]

THE END OF REDEMPTION: SLAVERY WITHOUT END

On January 9, 1442, shortly after Tristão's ships returned with seventeen captives, Eugene IV issued the bull *Etsis suscepti*, beginning with the obvious: "The . . . noble man Henry, Duke of Viseu, can rule over temporal matters and govern . . . [his] lands, possessions, and other movable and immovable goods whatsoever in the kingdoms and dominions of the King of Portugal." Continuing, the pope extended Henry's rule to include "any other places consistent with [the actions of] the military mentioned above [or] be granted by any [of the] faithful of Christ, or otherwise acquired by just means."

In other words, Prince Henry could claim ownership of lands and goods that his subjects acquired through any "just means," which included military actions or grants to him by any Christian. Since, in Iberian military tradition, the Crown nominally owned the goods and individuals seized, this decree included the men Tristão had kidnapped. Furthermore, according to the pope, he could hold these possessions as legitimately as he did in Portugal.[37] This resolved a potential objection to actions Henry had sanctioned. However, he faced a second, more serious challenge.

Prince Henry's endorsement of enslaving ordinary people, such as these fishermen, who went about their everyday activities, appears to have repeatedly met resistance. While almost no statements overtly criticizing his endorsement of such actions have survived, Henry undertook actions following their capture that could only have been prompted by doubts about the legitimacy of enslaving peaceful non-Muslims.

While Gonçalves and Tristão enthusiastically kidnapped and murdered defenseless people, some men under their command expressed reservations about the morality of their past participation in such actions. Prince Henry therefore asked the pope to issue a retroactive order of blanket absolution for anyone with moral qualms about their participation in these initial kidnappings and murders of non-Muslim innocents. Eugene IV obliged Henry, retroactively pardoning anyone with ethical reservations about participating in the seizures around Rio Douro. Issued only five days after the release of the bull in 1442, the order, called *Illius qui se pro divini*, stated, "Although at the time it did not matter personally, [to] soldiers . . . [who] waged war against . . . Saracens as mentioned earlier and *their [other] enemies* . . . [and] who have been contrite in heart and confessed verbally, we grant the forgiveness of [their] sins by Apostolic authority."[38]

Since the end of the eleventh century, popes had granted such blanket absolutions before military engagements with Muslims. However, in this decree, Eugene IV retroactively forgave sins committed the previous year. To counter any opposition to this legitimation of the slave trade, Prince Henry added an unusual postscript to the decree signed by Pope

Eugene IV, writing, "Therefore, it is *unlawful* for any man to violate this page of our grant [to seize Muslims and other unarmed people] *or to oppose it* with foolhardy recklessness." Henry's unmistakably threatening language effectively proclaimed his intent to silence current or future critics of the slave trade.[39]

Having secured an uncontestable moral right to the developing slave trade, Prince Henry authorized the single largest expedition yet to the bay that Tristão had tentatively explored two years earlier. Given the latter's grievances about the cramped quarters on his vessel, new ships were designed to more than double the cargo space. The larger hulls, in turn, meant that larger sails were needed to catch more wind to propel the greater weight of the ships. Thus, a significant improvement in the design of sailing ships allowed the vessels to carry more enslaved people.[40] Financing the bulk of the construction of these larger vessels was a wealthy resident of Lagos named Lançarote. Together with five other slightly less affluent men, he built a fleet of six larger ships to travel back to the area south of the White Cape. Each caravel had its captain, with Lançarote in the lead, followed by Gil Eanes (who had first rounded Bojador), Stevão Afonso, Rodrigo Alvarez, João Dias Amador, and João Bernaldez.[41]

Upon reaching the Bay of Arguin in mid-June 1444, the ships spent six weeks stopping at various places along the edge of the bay and on the sandy islands that lay close to shore, where the crews rounded up everyone they could find, from women carrying turtles across the beach to a young girl asleep in bed in a town where everyone else was out. As they proceeded along nearly four hundred miles of coast, they made their two largest captures via unprovoked military attacks. In one case, the Portuguese took only the smallest captives, because the larger adults would not fit in the landing boat. Altogether, they took around 240 men, women, and children, 232 of whom survived the month-long trip back to Portugal.[42]

By the end of July 1444, this large-scale expedition had seized those 240 individuals from fishing settlements along the giant bay. Following the Iberian Muslim and Christian conquest tradition, any captured booty (including human beings) had to be divided into five equal parts, the ruler receiving one-fifth. For some reason, the decision was taken to hold this apportionment in public on August 8, in a field outside the city gates of Lagos. But many people found both the sorrow among the captives and Prince Henry's brutality during the division of captured people repugnant, as even Henry's official chronicler makes clear.[43] Zurara wrote:

> How hard would a heart have to be not to be moved by compassion, seeing that group? Some hung their heads low, their faces bathed with tears seeing one another, others moaning very painfully, admonishing the height of the skies, fixing their eyes toward them, shouting as if asking for help from the father of nature. Others buried their faces in their palms launching their full length onto the ground. Others lamented in a song according to the custom of their land. Our people could not understand the words because they did not know the language well, [but] it [the song] corresponded to the degree of their sadness.[44]

But the real horrors were yet to come. Both Christian and Muslim Iberian traditions had long sanctioned the allocation of one-fifth of captives to the secular or religious leader. This goal could have been accomplished by selecting groups of parents and children or husbands and wives, thereby keeping families together. However, on August 8, Prince Henry's men undertook this apportionment ruthlessly by creating five equally sized groups so that each captain would receive equal financial benefit. This made it "necessary to separate children from parents, wives from husbands, [and] brothers from sisters. They [the men separating the enslaved people] observed no rules in separating friends or relatives; everyone landed where fortune had placed them."[45] In short, Prince Henry placed economic gain above even a modest gesture of humanity.

In vain, Henry's official chronicler hoped that "the workers of this division take pity into consideration over so much misery, seeing that some held on to one another so tightly that they could barely be separated."[46] He continued: "Children seeing their parents in another [group] jumped up and ran towards them, mothers grabbed their children in their arms and threw themselves on top of them, and without caring that they were injured, beaten or whipped for refusing to release them."[47]

Zurara was not the only one appalled by the actions he witnessed. The site was filled with people from nearby towns and regions who had come "solely to watch the novel [spectacle]. . . . Some left crying and others made such a noise that it disturbed the men in charge of the separations."[48] Prince Henry rode in, mounted on a powerful horse and surrounded by his men, whereupon he forcibly separated the forty-six people that comprised his share. The prince deliberately severed the bonds between captives by arbitrarily giving away some enslaved people to his men, "like a man of his status would do with a small treasure under his control."[49]

After many paragraphs detailing his revulsion over the horrors he had just witnessed, Henry's chronicler retreated from his emotional outburst—to instead justify Henry's brutality and to manufacture some redeeming value from the events by adding to the end of this account that the prince was pleased by the salvation of so many lost souls. However, there is no evidence that Henry had such motivations. The prince provided neither written nor spoken statements of such intentions. Instead, "services" to the prince continued to mean the obvious financial ones, just as serving God meant monetary gain.[50] Another eleven years would pass before the pseudo-rationale of conversion appeared in justifications for slaving expeditions.

By the start of the 1450s, these conquests had become the most lucrative activity for Portuguese sailors along the West African coast, amounting to more than a thousand people a year. Although their stated goal in capturing enslaved people was "to serve God and the Prince," it soon became apparent that "service" to the Almighty meant capturing people who would become enslaved.[51] After the largest raid of the era, Zurara added that God would "reward and pay for their labors and expenses in the work that they had undertaken for His Service on that day."[52] Thus, they could prove they were serving God and were in His favor by making a great deal of money by enslaving others.[53]

However, two uncertainties still hung over these large slave raids. First, the Iberian customs governing conquest, which stood for over a century, had only ever permitted enslaving Muslims. Since Muslims were circumcised, the hundreds of uncircumcised men sold in Lisbon over the next eight years did not, in the eyes of the Portuguese, follow the Prophet Mohammed. As a result, Prince Henry faced objections that he lacked the moral license to sell to non-Muslims. Again, he sought the imprimatur of the pope.[54]

At the request of Prince Henry and his uncle, King Afonso V (r. 1438–81), Pope Nicholas V endorsed enslaving three new groups in 1452: "pagans," "infidels," and "enemies of Christ." Because of "the fury of the enemies of Christ . . . it is valid for the Christian religion to subjugate [such people]. . . . [and] strive to overcome these enemies with a powerful arm."[55] At the stroke of a pen, in the bull *Dum diversas*, the pope made the very act of resisting enslavement into a justification for the same, and redefined the slave raiders' attacks as a struggle for God. Thus, purchasers in Lisbon might deceive themselves that they were acquiring individuals engaged in a vicious anti-Christian war because the prince, the king, and the pope had all certified people belonging to an unknown or unidentifiable faith as "enemies of Christ."[56]

Having established his sole right to rule over territories south of the Bulge and threatened anyone who opposed the slave raids, Prince Henry turned to end the historically valid path to freedom. Someone residing in Africa could still amass sufficient wealth (goods or enslaved people) to redeem a relative who had previously been transported to Lisbon. Adahu successfully made such an exchange in 1441, following the first raid. Yet, by 1452, no such options were available to the West African captives. To keep their share of profits from the sales of enslaved people, the prince and the king needed to prevent such redemptions in the future, putting an end to traditional options for ransom. All sales had to be final, and enslavement had to become irreversible. Thus, Prince Henry and King Afonso coldheartedly requested that slavery become permanent.

In 1452, Pope Nicholas V duly granted Afonso and Henry's brutal request "to reduce . . . Saracens, pagans, infidels, and enemies of Christ to perpetual bondage."[57] However, despite the pope's endorsement, Henry and Afonso's carefully drafted extension of the twin features of traditional conquest proved unsatisfactory. As the proclamation was formulated, purchase and exchange rapidly replaced war as the principal means of acquiring enslaved people. As early as 1444, Henry had indirectly supplied over twenty-five hundred meters of cloth to trade at Rio Douro in return for bondspeople.[58] Furthermore, raiding for ransom (and the possibility of freedom) continued in both Iberia and North Africa. Perpetual servitude as a punishment simply for nonbelief proved unconvincing despite the papal decree. Henry and Afonso needed a different strategy.

Thus, they approached Rome with a novel idea two years later, arguing that the journeys to West Africa's coast were missionary ventures that required perpetual slavery to convert these people to Catholicism. This pretext would prove to be their most successful rationale yet, popular among Portuguese slavers in the fifteenth century and adopted outside of Portugal by other Christians in the colonial era to come. This pseudo-pious justification allowed slaveholders to insist on self-styled religious motivations to justify slavery—since, for them, as for Prince Henry, enslaved people had become highly lucrative. Of course, even this rationalization seems feeble: Once enslaved people had converted to Christianity, they should have been freed, since the original legal justification for their enslavement no longer held. However, because the underlying motives were economic rather than spiritual, slavery's missionary logic remained unquestioned.[59]

To justify this new missionary narrative, Henry had to revise the history of the voyages past Bojador to accent conversion as his primary goal. As part of his rewriting, Henry pointed out that in October 1443, nine years after the initial search for local information had failed, he began to order nautical charts of the newly found coasts to be drawn up. "No one in Christian [territory] knew anything about this place, whether it was populated or not, nor was it even drawn on nautical charts nor in world maps."[60] By emphasizing the start of new mapping, Henry drew attention to his narrative of novelty. Following this start, scores of subsequent maps created the first accurate outline of Africa, which was completed sometime before the end of the century.[61] Tragically, nearly all these maps vanished in the tsunami, leaving no trace of how cartographers working for the Portuguese created techniques for more accurate coastal maps of Africa, leaving the only surviving cartographic record in the hands of plagiarists and thieves.

Henry began by emphasizing the novelty of the voyages he had sponsored or authorized, a revisionist approach to even recent history. By 1454, he had become aware that no previous sea voyages had occurred along the west coast of the Sahara, as he initially believed. So, he asserted that these "southern and northern regions of the Ocean Sea [the Atlantic] had never, or at least during the memory of mankind, ever been navigated, [thereby] making it so unknown to us to the West, that we had no certain knowledge of the people of those areas." Lacking prior knowledge of these people, he could rationalize Portuguese presence there on the grounds that they sought to have "the most glorious name of the same Creator . . . be extolled and revered even in the most remote and unknown places."[62] Henry wrote:

> By the praiseworthy works and industry of the said Prince, inhabitants of the various islands existing in the said [Atlantic] Sea, have come to the knowledge of the true God, have received sacred Baptism, to His praise and glory, and the salvation of many souls, [and to] the spread of the Orthodox Faith and the growth of Divine worship.[63]

Thus, the bare minimum proved satisfactory. Baptism—uttering incantations while pouring water over the heads of enslaved people—sufficed to pronounce a successful conversion. Once they learned Portuguese, they would be baptized, and their children and grandchildren would be born Christian. Further efforts to teach them prayers or Catholic doctrine appeared to be unnecessary.[64]

This reformulation shifted the meaning of the word *redemption* in Portuguese. The ongoing religious wars in North Africa continued to signify a prisoner's freedom in exchange for money, goods, or other people. But when referring to Africans from the Sahara southward, *redemption* became the pretext that slavery redeemed people for their heathen beliefs. Thus, in an Orwellian moment, slavery came to mean liberation.

This shift also required slave traders officially to justify future seizures as acts of "salvation." On Tristão's voyage in 1441 and Lançarote's in 1443, Zurara wrote that some went "to serve others, to gain honor, others with the hope of profit," which he declared amounted to the same goal—money *was* honor. Addressing slave raiders, this new directive allowed them to continue "only with the holy intention that you are acting to seek salvation for lost souls." Thus, slavers could seize as many people as possible, provided they declared they were observing the official ideology of religious redemption. In other words, the legal, religious niceties had to be formally stated, regardless of any underlying economic motives. Zurara concluded, "If their bodies are placed in some minor subjection, that is such a little matter for the help of their souls."[65] Forced and unpaid labor in a remote and unfamiliar land was thus reduced to "some minor subjection."

Although often wrongly described as general expressions of Catholic Church principles, all three preceding bulls (*Dum diversas, Etsis suscepti*, and *Illius qui se pro divini*) had formally addressed only Prince Henry and/or King Afonso, as did *Romanus Pontifex*, Pope Nicholas V's more expansive grant in 1455.[66] They never granted a blanket authorization to all Europeans to take people from Africa, nor did they imply such a prerogative. Portuguese rulers were keen to maintain their monopoly on slavery and its income. Hence, successive popes clearly empowered only the king and prince (and their subjects) to enslave "infidels permanently."[67] Thus, subsequent claims by other European powers that these decrees legitimated slavery generally overstate the case; it had only been approved for the Portuguese royal family.

Although some writers have employed *crusade* to describe these actions, that word is missing from contemporary documents in either Portuguese or Latin. *Croisere*, an early term for "crusade," came to be used in the thirteenth century to describe the war against the Islamic community in the Holy Lands; it was replaced by *crusade* in the following century. However, fifteenth-century Portuguese slavers had long sought anyone, be they Muslim, heathen, pagan, or infidel, who could be seized or traded from the coasts of Saharan and sub-Saharan Africa. Hence, in fifteenth-century Portuguese, *crusade* referred to wars against distant Muslims in the Holy Land; their actions in Africa were invariably called *conquests*.[68]

Prince Henry's role during this period alternated between two extremes. He persistently encouraged expedition after expedition until pilots figured out the path of the Long Return, thus gaining an understanding of the Azores High and the currents of the North Atlantic. However, beneath a superficial cloak of piety, he concealed genuinely vicious behavior, transforming a traditional exchange of prisoners of war into the permanent enslavement of Africans.

Although Zurara opened his account by lauding the prince, his powerful ending critique of Henry's behavior during the first slave market appears to be responsible for the book's disappearance from Portuguese archives (it was located in France during the nineteenth century).[69] Prince Henry exhibited extraordinary emotional detachment as the first large contingent of African couples and families wept and cried at being forcibly separated by a mounted prince in pursuit of profits from the slave trade. Yes, of course, they were humans, but to Henry their pain and grief were of no consequence—revealing an immense deficit of empathy that even his appointed chronicler could only report painfully.

4

How Trading Replaced Conquest

1444–1460

As the voyages sponsored by Portugal ventured southward, the ships passed two more capes named for their distinctive colors. The group Prince Henry viciously divided in the slave market had come from the first of these, the Bay of Arguin just south of the White Cape, whose shallow waters reflected the light desert sands. Next came the lush Green Cape, marking the westernmost point of the African mainland. Farther south the Red Cape awaited, named for its sandy, red-tinted water. Each cape, with its unique natural features—changing from near-white sand, to evergreen forest, to swampland beyond the red hills—demonstrated how the ecology of the West African coast continued shifting along with latitude. And as the Saharan sands ended, the types of people the Portuguese encountered in cultures residing along the Atlantic's coastal zone also changed. They inhabited varied ecological zones, ate different foods, lived in housing adapted to their environment, observed diverse rituals, and spoke many languages.

In 1446, only two years after Henry's cruelty in Lagos, a swamp-based Indigenous community south of the Green Cape launched a deadly counterattack on would-be Portuguese raiders, killing nearly all Portuguese crew and leading the Crown to rethink its methods of approaching new communities. Therefore, new ventures authorized by Henry searched for a means to replace the dangerous rationale of conquest as the central justification for onward journeys. After experimenting with sometimes absurd policies for approaching coastal communities,

Portuguese leaders eventually settled on a new set of protocols they complacently called "peace" instead of war. As a captain later claimed, "We have come here intending to treat the people with peace and goodwill, gaining friendship with them, which should be acquired with skill and not achieved by arms or force."[1]

However, this new "peace" strategy meant transferring the burden of capturing people for sale from the Portuguese to the local communities, thus freeing Europeans from direct participation in kidnapping—their role thenceforth would be merely "trade." However, even before they practiced "peace," they began constructing what would become a massive commercial system designed primarily for the exchange and selling of human beings, starting with the construction of the first warehouse (*feitora*) along the African coast.

THE WHITE CAPE: ESTABLISHING MORE PERMANENT TRADE

Prince Henry laid the foundations for this new system south of the White Cape, where residents of the area soon discovered that Portuguese visitors were willing to exchange goods for enslaved people captured and brought to the coast. At first, Arabs and the poorer Zenagas began to bring captives from the large Wolof and Mandingo communities nearby.[2] However, since these traders arrived on the coast at different times, not coordinated with the impending arrivals of Portuguese ships, Henry commanded that two fortified warehouse buildings be constructed near the White Cape, to serve as storehouses and to temporarily imprison people brought to the coast.[3] One of the makeshift detention centers was built at the mouth of a river just north of the cape, and the second was on the island of Arguin, just southeast of the cape and with easy access to ocean routes.[4] These storehouses served as temporary internment facilities that held people until they could be shifted to more permanent detention in remote lands—in short, they were designed for human trafficking. During the seventy years after its construction, the building on Arguin held over a thousand human beings per year before their transfer to Portugal.[5] The other warehouse, at the mouth of the river, failed due to a lack of trade, but the one on Arguin later attracted gold and salt caravans that passed nearby along their traditional route northward. The warehouse endured for two centuries as they began diverting their traffic to Arguin.[6]

Yet the ideology of conquest remained in the vocabulary and behavior of Portuguese sailors during their trips south from the White Cape, turning those trips to the Green Cape into a mix of conquest and trade. The Portuguese had yet to be outwitted by an Indigenous community. The time to truly renounce conquest and adopt Henry's warehouse-jail concept had not yet arrived.

Among the items the Portuguese took to trade at this first successful warehouse were djellabas, loose-fitting, ankle-length hooded robes worn by North African men; hambels, a type of rug for rolling up and sleeping under extensive lengths of cotton or linen cloth; scarves; and sometimes North African wheat. In return, they received gold, antelope hides, and an outstanding source of gum arabic to use in ink and paint, from the bark of the acacia trees whose dark green leaves gave the Green Cape its name.[7] Well into the next century, kings would construct similar storehouses along the African coast with the same dual functions. As with the warehouse to Arguin's north, some proved poorly connected to native trade routes and failed.[8] At least two such warehouses were built and soon deserted along the Guinea coast. But despite the mix of successes and failures, the practice continued as the Portuguese empire spread from West Africa to India and China, eventually creating a web of warehouses linked by the sea.

FROM WHITE CAPE TO GREEN CAPE

Unsatisfied even with the lucrative trade around the White Cape, Prince Henry soon endorsed two voyages south of the Bay of Arguin, following the currents. Seasonal streams allowed boats to travel to and from areas beyond the White Cape; from autumn through winter and into spring, coastal waters flowed directly south as far as the Green Cape. During a short interval from May to July, the current reversed and headed north at a slow 0.583 knots. While this pattern dictated an annual window for vessels' returns home, a second option was presented by the Mauretania Current, which allowed ships to return during autumn.[9]

Exploiting these new opportunities, Portuguese ships sailed past another two hundred miles of uninhabited dunes before encountering the mouth of a wide river the locals called Senaga.[10] Summertime provided the smoothest sailing for trading at the mouth of the Senegal River as currents flowed gently north and south, making crossing the river's wide entrance easy. However, the Portuguese could not extend their trade network farther inland, due to the number of waterfalls upriver.

On the north side of the Senegal River, Portuguese crew members laid their goods on the ground before the Indigenous community, who took them without uttering a single word and without reciprocating with their own goods. After crossing to the other side of the river, the Portuguese again laid out their cargo. Unlike the silent north-side residents, the south-shore natives immediately began speaking. Although no one from the Portuguese ship understood a word, they began to trade by using sign language and pantomime.[11] Eventually, it emerged that poorer Arabs resided on the northern side of the Senegal River, while a prosperous Wolof kingdom lay on the southern side.[12]

Wolofs had established a sizable inland empire connected to the trade in salt and slaves for at least a century before the Portuguese arrived. Thus, having long participated in trading rituals and exchanges with multiple communities and customs, Wolof residents at the mouth of the Senegal River readily adapted to trade with strangers.[13] Having access to communities in the interior by way of the river or on horseback, this coastal community began trading regularly with the Portuguese, exchanging enslaved people for prized Iberian horses, which were crucial to the slave trade but susceptible to parasitic infections.[14] Sometime after this trade began, the local chief gained enough wealth to become independent from the principal Wolof leader, who resided in the interior.[15]

THE GREEN CAPE

Having established lucrative trade on the shores of the White Cape, the Portuguese Crown continued sponsoring expeditions even farther south in the early 1440s. Leaving the Senegal River and traveling southward proved fastest during winter and spring, when the currents near the coast switched to a southerly direction, at a leisurely three-quarters of a mile per hour.[16]

Continuing for ninety-two miles south of the Senegal River, followed by three hundred miles of Saharan desert beyond the White Cape, sailors spied a vast, unexpected stand of evergreen trees with thick, rounded tops—the acacias, rising as tall as sixty-six feet (twenty meters).[17] Having imagined the entire area south of the Bulging Cape as a desert, Portuguese sailors were stunned to see giant masses of such trees without a massive river to feed them, so they named it the Green Cape (Cape Verde).[18]

The most notable aspects of this cape lay in its geography, rain-furnished landscape, and extension into the Atlantic. There would be no need to plant pillars to indicate a coastal landmark, as it would become necessary to do farther down the coast later in the century. The unforgettable deep greens emerged suddenly, directly in front of the southbound ships, clearly marking a sharp environmental contrast.

The first expedition to reach the Green Cape in the early 1440s returned quickly.[19] A year later, a second expedition took slightly longer, as its leader, Álvaro Fernandes, needed to anticipate how differently these communities would respond to the arrival of strangers. Heading slightly westward to round the cape, the caravel came across the island now called Gorée, inhabited by wild goats. A pair of canoes approached Fernandes's ship while his men slaughtered the goats and collected their meat. Fernandes welcomed the ten visitors aboard and allowed them to depart after giving them food and drink. When a more extensive group of thirty men appeared in canoes shortly after, Fernandes returned to the military tradition and ambushed them before they reached the boat. The crew pounced on the first canoe as the rest retreated. However, three men in this canoe swam safely to shore, leaving only two captured, and only after considerable effort. After sailing a few unimpressive miles farther south, Fernandes returned to Portugal.[20]

As a harbinger of what would come later, Fernandes had encountered considerable opposition from an adept deepwater maritime community with large boats and considerable swimming skills that few Portuguese possessed. However, Fernandes did not leave the island before indulging the impulse to mark his arrival by carving Prince Henry's motto, "The Talent to Do Well," into a tree.[21] Such skilled natives and deepwater canoes would become a recurring obstacle as other Portuguese expeditions traveled beyond the Green Cape, where the coastal landscape and the Indigenous communities would change.

GRAND RIVERS, SWAMPS, AND MANGROVES: FROM THE GREEN TO THE RED CAPE

Beginning around sixty-five nautical miles south of the Green Cape (now Dakar), Portuguese sailors noticed a difference in the coast. Instead of sandy beaches backed by dark green hills, a series of marshes came into sight. The appearance of this new coast would have been alien to sailors who had only ever traveled around Europe's Mediterranean or western coasts. Instead of a clearly defined shoreline with rivers easily visible from the deck of a ship, this coast gave the impression of widely separated clumps of mud, each held together by unfamiliar trees or grasses.

The trees anchoring these muddy stands were mangroves: tall, slender trees with many roots raised above the water to absorb oxygen. As ocean tides flow from one direction and river waters course from the other, clumps of silt and sand cling to these trees. Into this mucky silt came salt grasses, fishes, oysters clinging to mangrove roots, and birds resting on their branches, eventually securing entire swampy islands. Although forming only 0.5 percent of the world's seabed, these unusual plants store at least half of all the carbon in the ocean, earning the nickname "blue carbon." Of course, fifteenth-century sailors on Portuguese caravels would not have known this, nor would they have seen this unusual landscape up close.[22]

The tallest mangroves grew closest to the ocean, their height and dense branches blocking an approaching ship's view of the shore, lending no clue as to what lay behind each muddy clump. From the air, these irregularly shaped wetlands appear strewn across the deltas, resembling pieces of chopped vegetables being washed down by the water streaming from the rivers. As the fresh water streams out, salt water pours in, as if continually rinsing these islands in first one solution, then another.

Beyond these islands lie multiple streams of water, including tributaries either rushing to join with others or simply adding to the volume of powerful torrents from deep in the interior. It seems impossible to know the origin of each freshwater stream when they are viewed from the sea, since each one flows around many different marsh islands. Aboard their ships, sailors could distinguish some of the currents by noting differences in the clarity or muddiness of the water, or by feeling the motions of the boats (sailing being partly a kinesthetic experience). These coastal mangrove forests and tidal salt marshes extended nearly two thousand miles south from the Sine-Saloum delta and included twenty-one significant rivers.[23]

A DEADLY ENCOUNTER

The first Europeans to reach this region failed to anticipate that the people who inhabited it would have adjusted their way of life in order to survive and flourish among these marshy islands. Furthermore, they did not yet grasp the fact that these peoples' food-gathering and self-defense customs and practices would differ considerably from those of the more familiar forest and desert dwellers to the north. Any cautionary lessons that might have been gathered from Fernandes's encounter with Gorée remained unlearned.

In 1446, the first Portuguese ship to sail beyond the Green Cape passed miles upon miles of sandy beaches, backed by scrubland, before finally stumbling upon the broadest waterways anyone had encountered since the rounding of Bojador. Unlike the bone-dry Sahara, the Sine-Saloum delta was waterlogged. Miles of wetlands and scores of birds, turtles, and fishes confronted the sailors. As was customary, the ship anchored offshore near the mouth of one of the estuaries while the crew scrambled into two rowboats to explore the waterway. As a robust incoming tide carried their boats inland, they spotted a small settlement.

The powerful tide initially bore the rowboats past a dozen canoes carrying seventy or eighty locals waiting for them in the river. The Serer, who had resided in the region for several centuries, attacked the men in the boats with poisoned arrows made from a potent mixture of local plants, including one unique to tropical regions. Despite rowing furiously back to the ship anchored offshore, four Portuguese perished instantly in the rowboats, victims of the infamously fast-acting hydrogen cyanide.[24] Only two men managed to clamber aboard the ship, where they tried to save themselves by struggling to raise the ship's anchor caught in the mud. Serer arrows soon grazed both men, forcing them to cut the mooring, setting their rowboat adrift and abandoning those in the other rowboats, all of whom soon died.[25]

In the end, all senior members of the expedition perished, including the captain, Nuño Tristão. With all the experienced crew members and leaders dead, the ship was left with five unprepared young people to sail it: Aires Tinoco (a notary), an unnamed young African, two young cabin boys, and a single seaman. This untrained group labored for two months before successfully returning to Portugal's coast, bearing the poisoned arrows that had landed on the ship, but no record of the coastline.

After doctors examined the arrows, the leaders of the next expedition chose to bring a traditional Portuguese antidote to poison: urine and oil mixed with unnamed herbs. Upon returning to explore upriver of the Sine-Saloum, the captain, Álvaro Fernandes, was wounded by a spear with a poisoned tip. By immediately pouring the urine and oil over the wound, he somewhat reduced his dose of hydrogen cyanide, leaving him incapacitated but not dead. However, the injury's deterrent effects were apparent: vision loss, inability to walk, confusion, restlessness, dizziness, memory loss, unconsciousness, abnormal heart rate, and short or irregular breathing. Fernandes's ship returned to Lisbon, his fate never mentioned in surviving records.[26]

Having launched two such disastrous attempts at contact, a third sortie would proceed more cautiously but still not carefully enough. In 1448, a nine-caravel fleet sailed from Lisbon toward the Sine-Saloum

MAP 4. The West African coast from Sine-Saloum to the Casamance River. NASA.

delta. Although stopping short of Tristão's expedition, this venture's more tentative approach included sending a single crew member to scout the territory inland. Having discovered cotton and rice fields, he rushed back to the ships, where he gathered a group of fifteen to accompany him inland. However, in his excitement over this discovery, he failed to notice Serer lurking nearby. The fifteen promptly set off in the direction of these fields, where they were met with a hail of hydrogen cyanide–poisoned spears and arrows that promptly killed five and injured another two. When a second group saw the fate of the first, they hastily retreated to their ships, where they weighed anchor and fled northward to the well-known waters of the White Cape.[27]

Chastened by their experience with the Saloum River natives, the survivors of this fleet remained cautious. Rather than heading off into the interior, as Antão Gonçalves and others had done, they interrogated friendly Muslims along the coast about the locations and activities of local Serer villages before setting off into the interior. Targeting peaceful, unarmed villagers or enslaving fisherfolk who had never done them any harm would be far safer. Such actions met with the approval of local Muslims, who have continued to enslave the Azengues (Haratin) to the present.[28]

After this third disaster, Lisbon officials realized they needed to stop sending their people into the interior. In the words recorded by a chronicler reflecting the apparent command, "subsequent efforts would not proceed with as much intensity and vigor as previously [but] from this year onwards [1448], actions in those places [Africa] should always be *more by trade and an adequate quantity of merchandise* than by exertion or force of arms."[29]

In the future, Portuguese and Italian accounts would describe these self-preserving instructions as establishing peaceful commercial relationships in Africa. However, in one recorded exchange, the natives vehemently disagreed.[30] In 1455, Portuguese sailors met with Gambia River dwellers who believed that Christians ate human flesh and purchased Africans only to eat them. This interpretation makes sense, given that the locals saw enslaved people being placed in the hold of (swallowed up by) a giant ship, never to be seen again. Although the Portuguese did not eat the flesh of Africans, they did use and dispose of them with little regard for their humanity, so perhaps the difference is academic. After this decidedly hostile response, the crew refused to continue upriver and returned to Lisbon.[31]

For the remainder of the century, Portuguese leaders instructed their expeditions to retreat if they encountered armed natives massed in canoes, rather than invite conflict.[32] This way, the Portuguese could operate with a relatively small military force, avoiding needless casualties in clashes with natives and reserving their military resources for defense or protection of commercial activities.[33] While Prince Henry had reluctantly abandoned the direct seizure of people and goods, he struggled to find new approaches to making initial contact with previously unknown people and finding goods to exchange. This task proved far more challenging than he could anticipate, resulting in countless failures to develop an effective strategy. It would take seven years of failed trials to find the proper protocols for first contact.

FAILED TACTICS OF ENCOUNTER

One of the first such endeavors occurred during a 1448 voyage, mainly at a Danish nobleman's request to bring an entire elephant skeleton back to Denmark. (To this day, the Order of the Elephant, a prestigious Danish order of chivalry, remains the highest royal honor conferred in that country.) Prince Henry gave the nobleman a ship, a captain (Fernan Afonso), a pilot, an interpreter, and a crew. And, for the first time, the crew's instructions included a protocol for establishing peaceful contact.

The crew was also ordered to carry a written letter addressed to the ruler of the area north of the Sine-Saloum, asking him to guarantee peaceful trade. This diplomatic approach, while commendable, was based on a flawed assumption. The idea that a formal exchange of written letters could secure trade seems absurd in retrospect. It assumed that a text secured a relationship, that a single ruler exercised control over coastal contacts, and that he authorized commercial relationships. Furthermore, the letter arrogantly concluded with a demand that the ruler assist the Portuguese in fighting Muslims, reflecting a baseless assumption—one that would recur several times over the next fifty years—that he would necessarily be Christian because he was not Muslim.[34]

Ordered not to land until they had received a written response from the local ruler, the forewarned Portuguese remained offshore in their boats, their translators shouting to people on the shore. How or indeed whether they handed over the letter is unclear from surviving sources. However, the Danish nobleman, who was part of the expedition, lost patience as they waited and demanded to be rowed closer to land, ignoring the crew's warnings. Nearing the shore, he grabbed a gourd and struggled with the man holding it. The rowboat promptly tipped over, allowing the single rower who knew how to swim to return to the larger boat and safety. With no sign of the expected letter, the Portuguese left the Dane to struggle for the gourd, weighing anchor and sailing back to Lisbon. This tactic for connecting peacefully with African communities would not suffice.[35]

After the misadventure with the Dane, ships stayed north of the Sine-Saloum delta for the next seven years. Henry licensed independent journeys in 1451, 1453, and 1454 to the region around the White Cape. In 1451, one Abrão de Paredes was allowed to trade on the African coast; two years later, a pair of caravels owned by Nuño Antunes de Gooes were licensed to trade in springtime along the Rio Douro peninsula.[36] Coastal surveys of the area between the Sine-Saloum and Red Cape would have to be postponed until a strategy for close encounters with the Serer and related communities was developed.

During this period, the Portuguese Crown solicited participation from two traditional Mediterranean maritime regions. King Afonso V of Portugal signed a compact with the Duke of Brittany to allow Bretons experienced in Atlantic navigation to trade along the West African coast for six years, from 1452 to 1458. During the same decade, the Crown signed individual agreements with three Mediterranean mariners: Alvise Cadamosto, from Venice; and Antonitto Usodimare and Antonio da Noli, both from Genoa.[37] When encountering any group they believed might possess poisoned weapons, these captains beat a hasty retreat. "The fishermen of the region attacked me with bows and poisoned arrows, judging us to be enemies. And I, seeing that they did not want to welcome us, was obliged to return about 700 leagues," wrote Usodimare.[38]

These challenges pushed explorers to bypass dangerous ports and go farther south. Fifteen nautical miles to the south of the deadly Sine-Saloum, the Portuguese encountered what they initially called the "Great River." Eight land miles wide at its mouth and seven hundred miles long, it is now known by a more apt regional name, the Gambia.[39] In 1456, several ships arrived at this broad river after the Portuguese had adopted an exchange-based approach. The crew began by enticing a resident on board with displays of goods and then sending a valuable gift to his chief. These more openly commercial moves succeeded where previous brutal displays of force had failed.

The Venetian Cadamosto adopted this approach when he reached the large Gambia River, where he encountered several unarmed men in canoes.[40] To entice one or more of them aboard, the sailors displayed their goods where the natives could see them, much like riverside hawkers today on the Amazon, who dangle their wares on long poles they hoist alongside the riverboats. Despite the display, only a single man approached and boarded the ship. Knowing they would need someone to guide them to the local leader, the Portuguese crew gave their visitor presents, food, and accommodation. In return, he indicated the route upriver to a regional chief's domain and accompanied a man from the ship.[41]

Cadamosto's emissary carried a substantial gift: a loose-flowing overcoat of the finest silk that suitably impressed the chief.[42] He stated (in the chief's Mandinka language) that he had come on behalf of the Christian king of Portugal to establish a "good friendship," adding deceptively that the king annually would send the chief any Portuguese good that he desired. The offer feigned royal humility, declaring that the new arrivals wanted "to understand from him what he [the chief] needed of the things of our country."[43] The promise of future regular deliveries from Portugal gave the chief an unparalleled opportunity: He could gain a reputation for generosity or fill his coffers by becoming the intermediary for those who wished to acquire novel goods. However unlikely such future access might be, this meretricious promise ensured that the chief would allow the European visitors to acquire local goods without fear of losing their lives.

Over the following two weeks, the crew traded goods with the chief for enslaved people and a disappointing amount of gold. However, the scope of the trade gradually expanded. As the days passed, people from far and wide began appearing in canoes to sell a multitude of animals, including civets, whose parts were used in expensive perfumes,[44] apes, and parts of an elephant; however, only the cloth, dyed in red, white, blue, or stripes, struck the crew as promising for future trade.

Having concluded these successful exchanges at the Gambia, Cadamosto continued southward in search of commercial opportunities. After nearly sixty miles of sailing, he came across another large river and once again sent a party of interpreters ashore to inquire about the area, including the names of the river and the local ruler. This time, they returned quickly, saying that the ruler, Casamance, lived thirty miles upriver and was away fighting a war. The tale of the absent ruler, far upriver, combined with breakers extending four miles out, successfully deterred Cadamosto, who sailed off and proceeded farther south.[45]

Upon reaching a vast river, likely the Rio Grande de Buba, the boats anchored near its entrance.[46] The following day, two elongated canoes rowed swiftly toward them, one carrying thirty men and the other sixteen, each nearly as long as their own vessels. Suspecting an attack, the crews promptly armed themselves. However, as they neared the caravels, the rowers attached a white cloth to the end of an oar. Relieved, the Portuguese responded with an identical gesture and put aside their weapons. The two canoes then separated, with each coming alongside a single caravel. Although Cadamosto believed that the locals were unacquainted with multi-masted ships, the rowers' familiarity with a standard European convention (the white flag signaling peace) and their willingness to board suggests they had had previous engagements with Europeans.[47]

Sixteen miles south of the river where the absent chief resided, Cadamosto recorded a new sighting: the distinctive cliffs that form the Red Cape. He wrote that the cape was "higher than the land of the coast, the front of it being red, and for this reason we call it the red cape." This ordinary cape marks the beginning of a significant shift in the direction of the African coastline, as here it angles sharply southeast for the first time.[48] Despite this, the coastline becomes even more swampy: Occasionally, sandy beaches and green trees appear north of the cape between the marshy rivers; beyond the Red Cape, the coastline becomes one continual swamp.[49]

Finally the "missing" shoreline south of the Green Cape could be sketched without endangering Europeans. Moreover, this new, less violent first-contact strategy allowed pilots to survey the coastline without fearing poisoned arrows. Therefore, Cadamosto and others gathered information about the oceanfront south of the Green Cape, carefully noting all the major rivers where trade had been identified or might be found.[50]

Cadamosto continued beyond the Red Cape in 1456, proceeding quickly southeast as the rising tide eased the vessels' way for eighty-two miles.[51] The sandy beaches disappeared and he came across three additional rivers. The first, Saint Anne, was "a reasonable river . . . long" but with a narrow entrance. Access to the second, the Cacheu (São Domingos), was made difficult by the many breakers and sand spits. However, presenting few obstacles, the third river was even more expansive than the impressive Gambia.[52]

Cadamosto describes reaching "the mouth of a river so wide that in the beginning, we thought it was a gulf . . . [and] judged its width to be 20 miles at least." While not quite that broad in modern surveys, the entrance is still six times wider than the Gambia's. Despite difficulties navigating this gulf, Cadamosto spotted some islands and anchored offshore, without recording his exact location. Upon entering the gulf, he was surrounded by mangroves and many islands.[53]

On one side of the gulf lay a large cluster of islands, many inhabited. Although the full extent of this three-thousand-square-mile territory would not be known until the twentieth century, the Bijagos archipelago remains easily identifiable from 250 miles in space. Cadamosto dropped anchor somewhere in this vast expanse, not far from land, and likely entered the eighteen-mile-wide mouth of the Rio Grande de Buba; however, the politics of first contact reversed on this occasion, as Indigenous people initiated contact before the Europeans could.[54]

The morning after they anchored near the entrance to the Buba, when the locals met them and boarded the caravels, they began pantomiming the objects they wanted, and the Portuguese pointed to the locals' gold rings in response. However, to the Europeans' frustration, none of the multiple interpreters on either ship could understand a single word of Bak, their unique language.[55] The Portuguese decided to turn back.

Environmental characteristics of this enormous delta also hastened the decision to return. The day the first Bak speakers' canoes came alongside them, Cadamosto's ships struggled to remain in place. Even after the sailors dropped a third anchor and continually adjusted the sails, the mighty tides roared twice as fast as they receded, moving the ships and shocking an indignant Cadamosto, who found the tides most uncivilized: "In this area, we found great contrariness. . . . The rising and ebbing tides do not act as they do in Venice or all the West." Indeed, the tides flowed swiftly, between two and three knots in the dry season and between five and six knots after rain. Cadamosto also bristled at the unequal duration of the tides, complaining that they took only four hours to rise but eight hours to retreat. When faced with losing an anchor on a second day, Cadamosto swiftly turned his ships around and sailed back to Lisbon.[56]

Unfamiliar tides had ended a previous voyage in the Bay of Arguin. Near the southern end of the bay, Gonçalo de Sintra and twelve of his men disembarked at low tide on the island of Naar (Nâyer). After quickly crossing two sandbanks at midnight, they decided to wait on the shore until the tides receded. When they awoke the following day, they discovered that instead of completely receding, the small whirlpools had remained in the wake of receding tides. These eddies forced the crew to stay on land, since only three could swim. With the sailors stranded, local inhabitants attacked the twelve, killing seven, including the captain. The Bay of Arguin thus taught the Portuguese a grim lesson: Tides and their eddies could prove unpredictable. However, after Cadamosto's experience with robust ebbs and flows, future expeditions would not be caught off guard by powerful and irregular tides.[57]

Notwithstanding his frustration with the tides and his failure to identify new trading partners, Cadamosto's voyage had turned up fresh information on the direction and nature of the coastline south of the Red Cape, news that generated interest back in Lisbon. Portuguese leaders learned of an entirely new direction to the African coast for the first time since they had rounded the Bulge twenty-two years earlier. From the Bulging Cape southward, the shores mainly lay north to south. However, the coast slanted diagonally toward the southeast past the Red Cape. This tack suggested a new opportunity: the chance that Portuguese ships might head eastward toward the Indian Ocean. When Cadamosto returned to Lisbon, this sharp shift in direction caught the attention of King Afonso V.

This new southeasterly coast provided concrete evidence for the first time that a route around Africa to India might be found. King Afonso wanted to know what lay ahead. At age twenty-five, he learned that his uncle, Prince Henry, then sixty-three, his energy flagging, could not direct the southward expeditions for many more years. So, after inquiring about the most knowledgeable mapmaker in Europe, he chose to settle upon Fra Mauro, a monk in an obscure community of hermits on the island of Murano, in the lagoon of Venice. Fra Mauro worked for three years on the map, incorporating all the knowledge he could amass about the areas east and south of the recently discovered coastline.[58]

Completed the year after Henry passed, the map proved an immense disappointment. Fra Mauro had incorporated summaries of all the texts he had read, but he possessed no more knowledge than anyone else about the shape of the coastline to the south. After receiving an update on the apparent existence of an inlet south of the area Cadamosto had explored, he drew a small bay, which he labeled the "Gulf of Ethiopia." He inserted a single curved line to the south, adopting the conventions of Arab cartographers who had settled along part of the East African coast and knew that an inland area existed but had never ventured far beyond the harbors.[59] After finding that the learned mapmaker possessed no more knowledge than he did, Afonso realized he would have to proceed with the voyages farther south with no indication of what lay in store. However, the king also knew he would need to adopt a more pragmatic approach to first contacts than his uncle's.

Central to these recently adopted tactics would be the ability to garner accurate information about the commercial opportunities offered by these unfamiliar cultures and the rituals of exchange they expected. To that end, the Portuguese Crown relied on a culturally unique approach to native languages—perhaps the only saving grace in a history of unmitigated cruelty interspersed with the innovation of unique nautical techniques.

5

Language

How Enslaved Interpreters Established Trade

Understanding the local languages was vital to Portuguese relations with African communities. Customarily, Europeans learned only the smattering of phrases needed to trade, a pattern the Portuguese initially followed, ordering their enslaved interpreters to learn the local names of necessities only. However, unlike the Spanish, who would later insist in the Americas that natives learn only the Spanish language, the Portuguese eventually sought to acquire more expansive Indigenous vocabularies. After 1450, enslaved translators served an additional function as the Portuguese began acquiring goods produced by these local economies. Since their empire's economic prosperity rested upon mutually satisfactory exchange with native producers and traders, clear communication of shared expectations was crucial. Their relatively broader understanding of native discourse, therefore, did not signify that the Portuguese treated natives better or valued them more; rather, their exchange empire depended on a grasp of subtleties of language, which they recorded in the Lisbon archives. Thus, the loss of records during the 1755 earthquake deprived us of vast linguistic and cultural knowledge from fifteenth-century Africans.[1]

Portuguese travelers failed to contact any other peoples for five years after learning how to round the Bulging Cape. Only in 1441 did they come across a group of people in Rio Douro with whom they expected to speak—an expectation that was soon confounded. It is important to recall that Portuguese people came from multilingual worlds. Much of the kingdom was bilingual; in addition to Portuguese, tradespeople spoke Arabic in the markets and on the street, and diplomats in the royal court spoke more formal Arabic.[2] Therefore, the leaders of the first successful raiding party assumed that someone in the crew would be able to speak and understand the local language. Indeed, four crew members had a basic knowledge of Arabic, likely in the same way that people living in the southwestern United States may know some Spanish phrases and those in the northeastern states grasp some Canadian French. However, of the four, only one crew member spoke Arabic fluently—an unnamed man described simply as a "humble" servant of Prince Henry, as recounted in chapter 3.

Betraying their assumption that the Arabic spoken in the Sahara would be identical to that heard on the Iberian Peninsula, expedition leaders Antão Gonçalves and Nuño Tristão discovered that all of their kidnapping victims, save one, spoke either a Berber language or highly Berber-inflected Arabic that ordinary crew members could not grasp. Only the elite captive Adahu spoke a more standard North African Arabic, making Prince Henry's servant the only one who understood him. As with any error that impeded trade, Portuguese captains would not repeat this mistake. Neither would Henry's courtiers, who grasped the importance of shared language for information on sailing conditions or the location of potential riches, including enslaved people. A concerted effort would have to be made to develop commercial and linguistic resources for the seamen.

Over the next three years, Portuguese raiders seized around five hundred African people; during their captivity in Portugal, many became fluent in Portuguese. As the time approached for the first large-scale expedition in search of more slaves, members of Henry's court interrogated these enslaved people, in order to choose the three most likely to be valuable as interpreters. The prince selected two as official translators, setting a tradition from this expedition until the end of the century. Again, though demonstrating an awareness of the distinctiveness of other languages, the judgments of the Portuguese in these matters possessed neither humanitarian intent nor sensibility.[3]

In addition to sending a captive interpreter with the expedition, Prince Henry also impressed on its leader, Lançarote, the importance of seizing a knowledgeable native who demonstrated the potential to become a translator and could transmit information to the Crown about the location of settlements, possible gold deposits, or other valuable resources that might be found in the region. In chronicler Zurara's words, Henry "wanted to learn news of this land." Moreover, the prince specifically wanted a Zenaga Berber speaker to become an interpreter on future voyages south of the White Cape.[4]

Lançarote clearly understood the value of translators, insisting on their presence during any attack. When he sent an advance party to scout out the exact location of the people residing on Naar (Nâyer), an island in the giant Bay of Arguin, south of the White Cape, the group hesitated and turned back, despite sighting the settlement, because they lacked an interpreter, which Lançarote had decreed indispensable. The following raid (accompanied by an interpreter) seized more people than any other attack during the trip. Despite recognizing the importance of his translators, Lançarote mistreated them. When a Berber-Arabic speaker led Lançarote to an abandoned site, the captain beat him until he confessed to knowing of another settlement, which Lançarote then raided.[5] A year later, the prince selected a young Portuguese-speaking Zenaga captive to accompany Gonçalo de Sintra's expedition. However, he abandoned the Portuguese at the first opportunity and fled back to his people.[6]

Later, in the 1440s, Prince Henry reluctantly abandoned the increasingly dangerous policy of direct seizures through the flamboyant military approach of conquest. However, he still needed to develop a more durable rationale for continuing the slave trade along the Atlantic coast. In words he drafted for the pope in 1453, he stressed his own steadfast and laborious efforts:

> With the greatest labors, dangers, and expenses, every year he [Henry] ceaselessly sent the swiftest ships, the caravels, to search the sea and the maritime provinces to the south; and . . . navigated further to the ports and seas . . . at the mouth of one of the great rivers.[7]

The unselfconsciously arrogant Henry took sole credit for his role in authorizing the voyages and seemed unwilling to consider that his subjects and his slaves endured the most severe labors and dangers. Henry had failed in his initial search for a better military approach, even drafting a nonsensical letter in 1448 demanding cooperation in a crusade. Henry next had turned to the measures described here, relying on the skills of interpreters who had been enslaved. Alvise Cadamosto recalled:

> All our ships carried black interpreters brought from Portugal . . . [who were] enslaved people that the chief of the Senega [River] sold to the first Portuguese [and who] became Christian in Portugal and learned the Hispanic language [Portuguese] very well. . . . By compensating their owners . . . [we acquired] a considerable number from all over to work as interpreters.[8]

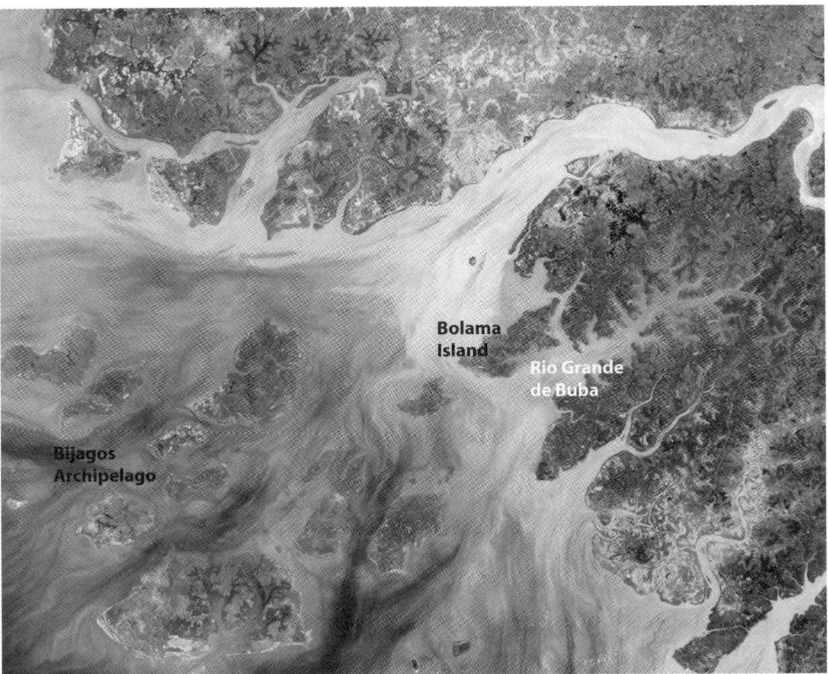

MAP 5. The Bijagos Islands, Bolama Island, and the Rio Grande de Buba. NASA.

Expedition leaders characterized their use of enslaved interpreters as benevolent, describing an improbable scenario in which each interpreter would acquire four enslaved people for his master and then be freed. Without any means of enforcing such a rule, enslaved people depended on their owners' goodwill to credit them with the captures and to free them.[9]

However, these enslaved interpreters lacked the autonomy to engage with other people, and often found themselves being treated as disposable. European crew members cast lots among themselves to decide whose slave would first set foot on unknown soil.[10] Thus, enslaved people lacked even the freedom to initiate contact. After disembarking the enslaved interpreter, rowers from the crew could withdraw safely offshore to await the results of the interaction. When natives murdered the translator, as they did along the Gambia in 1455, Europeans quickly fell back to the safety of the giant ship.[11]

Having sailed to the west coast of Africa for over twenty years, Portuguese slavers had acquired interpreters who spoke one or more of the two major language groups from the northern region: Mandingo and

Wolof. However, they needed help comprehending what turned out to be Bak, a language group unique to the waterways and offshore islands of modern Guinea-Bissau. Unable to establish a common vocabulary and a trading relationship, Cadamosto assumed that the languages farther down the coast would be equally alien and decided not to proceed.[12]

After Henry the Navigator died in 1460, King Afonso V, his nephew, succeeded him in overseeing the African journeys. As Portuguese voyages had proceeded near the Red Cape, they had begun to encounter communities like those near the Bijagos archipelago, who spoke languages other than the trio of Arabic, Wolof, and Mandingo that had served the Portuguese well before. This reality demanded a new policy. Afonso ruled out military action, as well as Henry's tactics, which had gravely endangered enslaved interpreters. Afonso's latest scheme, contained in orders to Pedro de Sintra,[13] sought "to take . . . blacks by peace or by love or by force [and] bring them to Portugal," where they would learn to become interpreters.[14]

TOWARD A NEW POLITICS OF ENCOUNTER

According to this retelling, Sintra's first attempt to understand an unfamiliar language occurred in the islands scattered amid the mangrove swamps of the Bijagos archipelago. To comprehend the still alien languages spoken on these islands, Sintra sent at least one enslaved interpreter ashore on an island, where he was welcomed. The interpreter visited homes, where he might have been able to find similar words for customary household objects. This determined effort to find a common vocabulary failed and the ships departed, no more enlightened about the Bak languages than when they had arrived.[15]

Following this failed attempt, Sintra began searching for yet unknown coastal communities farther south with which the Portuguese could trade. However, environmental and human obstacles prevented him from approaching the shore for most of his voyage beyond the Red Cape. The journey's most memorable moments occurred when his crew glimpsed dramatically shaped hillsides from the ship.[16] Less than eighty miles south of the archipelago, Sintra spied a peninsula with two large hills rising sharply from the low green coast. After rounding the point, the sailors on deck glimpsed the unusual sloping top of a hill nearly five hundred feet (148 meters) in height. According to the sailors, the slant resembled a diamond, though more recent nautical guides describe the shape more prosaically as a sloping roof. Despite being visible only from a distance because of the rocky shore, the dramatically shaped height of Cape Verga would prove helpful for verifying location at sea on future journeys.[17]

The expedition continued southeast of Cape Verga's distinctive contours but recorded nothing of the coast. Sixty miles later, the ships entered a large bay bordered by a long peninsula to the south. Once they were anchored in the Bay of Sangaréa, a flotilla of large canoes, each carrying thirty to forty men, came out to meet them. Surprisingly, their language was familiar to the Mandé-speaking enslaved interpreters on board.[18]

Two hundred years earlier, this group, the Susu, had fled a Mandé-speaking area, and they still spoke a dialect the interpreters could understand. However, the Portuguese found them a dim commercial prospect. The Susu farmed only subsistence crops, planting and harvesting without iron tools. Most wore ear and nose rings, while the elite men and women also owned gold piercings that ran through their lips. Although their goods lacked commercial value, the encounter proved peaceful and trouble free. However, one aspect of the environment near the Susu settlement would leave its mark on maps of the region for centuries.

The south side of the bay where the Susu resided extended for twenty-two land miles out to sea. Just barely one nautical mile beyond this headland lay six islands with two large semicircular islands that resembled opposite sides of a clamp, grabbing two tiny islands in the middle. The residents of these islands, whom the Susu called "people of the seaside," carried intricately carved wooden deities as they tended their rice fields. Since the Portuguese considered these carved images idols, their island residence became known as the Islands of the Idols. Later English and French navigators grasped only the *Los* part of the name *Los ilhas de ídolos*, and so these distinctive isles became the infamous "Los Islands," near the modern city of Conakry.[19]

Despite the lack of commercial opportunities with the Susu, the ships continued south from their enclave, hopeful of finding additional Mandé-speaking communities. However, this optimism soon evaporated. Just beyond the Islands of the Idols, on the southern side of the Kaloum peninsula, the expedition contacted a second group of subsistence farmers who spoke an entirely different language, likely Nalu, leaving the enslaved interpreters at a loss. Confronting artificial and environmental obstacles, the expedition failed to contact additional communities along the next hundred miles of coast.

Beyond the Kaloum peninsula and this Nalu subsistence community, the mangrove marshes and swamps reappeared. Since these environments

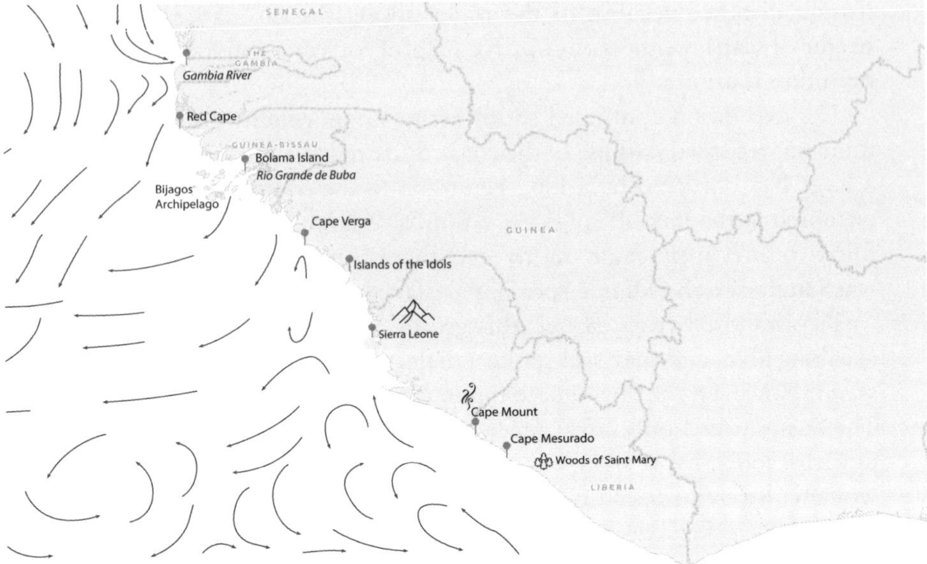

MAP 6. Pedro de Sintra's voyage. © Esri, Tomtom, Garmin, Open Street Map Contributors and GIS User Community.

resembled those where Tristão and his companions had met their deaths, Sintra stayed far from the coast. Nonetheless, his pilots did spot two of the six major rivers of the region, which they named the Saint Vincent and Green Rivers, respectively.[20]

Just over a hundred miles southeast of the sloping Cape Verga, Sintra encountered a second distinctively shaped mountainous peninsula that would become one of the most famous landmarks on the West African coast. Sintra named it Lion Mountain, but it is more commonly known as Sierra Leone. Sintra would later tell a fellow navigator that he named the mountain not for its shape, but for its ferociously steep hillsides. This still densely wooded mountain jutted twenty-five miles from the coast, ending with three islands lined up like an arrow pointing toward the peninsula's southern tip. Unfortunately for Sintra, no people resided near the shore or on nearby islands. Continuing his southward journey, even though he had successfully sailed past miles of coastal marshlands and mangroves, he continued to meet challenges from the environment.

Just thirty-eight nautical miles south of Sierra Leone, Sintra entered precarious straits where powerful tides (2.5–4.0 knots) shoved the sand in all directions with every rise and fall of the sea. Maps drawn after this voyage would soon display these treacherous sands as outsized tentacles stretching far into the Atlantic. Sintra successfully escaped these sandbanks' long furling and unfurling grasp by sailing some distance from the coast. Then, just as suddenly as they had appeared, the sandbanks vanished entirely once the ships reached a place that Sintra named Cape Saint Anne. Despite the easier sailing thereafter, Sintra could not contact any community for the next two hundred miles, due to two significant environmental hazards: When sailing beyond the sands, danger originated in the ocean; at the entrance to the first considerable river, some ninety miles south, the surf pounded so heavily that entering or even anchoring nearby proved impossible. Sintra soon encountered another environmental challenge, this one of human origin.

For the next 110 miles, the inhabitants were burning the forest along the coast to clear it for the new season's crops. This smoke also signaled to Sintra that the residents were fully occupied and, therefore, would lack the time or inclination to initiate contact with new visitors. Fire also posed an enormous danger to wooden ships like Sintra's. Eventually, the fumes became so intense that Sintra called the following major waterway he encountered the Smoky River.[21]

Smoke continued to cling to the coast as Sintra's caravels sailed past two prominent capes: first, the distinctive three-hundred-foot (ninety-meter) Cape Mount; and then, almost fifty miles on, a second, lower cape that he named Cortese or Mesurado. As they anchored off Cape Mesurado, they witnessed nighttime fires burning on the beach and in the woods. When morning came, they saw people moving about but postponed contact again, recognizing the potential risk from fires on the beach.[22]

The peninsula of a second, smaller cape extended another nine miles beyond its peak. But south of this peak, the fires abruptly ended as the crew members finally began to breathe clean air.[23] At the end of this heavily forested peninsula, Sintra's ships anchored at a place that he called the Woods of Saint Mary, now known as the False Cape. After a short time at anchor, a group of small canoes emerged. Like the Mandé-speaking Susu encountered earlier, the residents rowed out to meet the ships. This group's weapons seemed flimsy, and their initial attitude was more curious than hostile.[24]

As the flotilla of small canoes approached, Sintra's crew observed that each carried only two or three men armed with sharpened sticks and darts, leather daggers, and what appeared to be cleavers. All wore multiple earrings; some had nose piercings, and a few carried what seemed to be teeth in chin piercings. The Mandé and Wolof enslaved interpreters shouted to the men in the canoes but failed to understand a word of their response. However, Sintra had received unique orders for such encounters from King Afonso, guidelines to securing an economic advantage. His orders required him to abduct local interpreters.

A NEW STRATEGY TO FIND NATIVE INTERPRETERS

King Afonso's new directive empowered Sintra to entice or even kidnap individuals from a peaceful community that spoke an unfamiliar language. In the words of the briefing quoted earlier, Sintra was empowered to take "blacks by peace or by love or by force."[25] In short, he was to extract individuals by fair means or foul and return them to Portugal to become interpreters. The primary objective of initial contact had shifted from simply locating valuable trade goods to finding friendly coastal communities in possession of such goods and, just as importantly, amenable to supplying visiting Portuguese ships. Hence, this pragmatic approach, wielding both carrot and stick, aimed to secure the cooperation of coastal communities in extracting resources and transferring them to the Portuguese.

The initial application of these new measures occurred at this coastal community in the Woods of Saint Mary, at the end of the Cape Mesurado peninsula. Even though the residents lacked commercially valuable goods, Sintra's caravels had reached the farthest south that his supplies would let him sail. As the king would later note, taking an interpreter captive from the southernmost point on a journey could potentially provide the key to understanding other languages spoken farther down the coast. Hence, despite lacking valuable goods, Sintra followed orders to detain individuals as captives and future translators.[26]

Once the canoes reached Sintra's ship, the crew helped its three men on board. After some time had elapsed, Sintra let two men return to their canoe while he detained a third. It is impossible to determine whether this individual was kidnapped or agreed in some manner to accompany the ship back to Lisbon. In any case, if he had proved uncooperative, the detainee might have been auctioned into slavery. However, since he appeared willing, he was treated courteously.

The goal of translation operated in both directions. Although the Portuguese wanted to teach the natives their own tongue, they also wanted their translators and brokers to learn enough of the local ones to conduct trade. Quoting King Afonso, this extraction had "the goal of [first, the Portuguese] understanding [this captive] over time or [second,] finding a fellow speaker among the many other black translators who resided in Portugal or [third, the captive's] learning to speak [Portuguese]."[27]

Thus, upon Sintra's return to Lisbon, the new arrival from the forest of Cape Mesurado was placed into the care of royal officials, who were instructed to carry out a program that began with locating another speaker of his language. At first, Portuguese officials attempted to locate anyone in Lisbon who spoke a mutually intelligible language. Despite presenting him to every African interpreter in Lisbon, the king's courtiers failed to find anyone who understood the newcomer. Second, the courtiers began the laborious process of accompanying the man to the homes of all the city's slave owners. After countless introductions, they finally stumbled upon one enslaved person, a woman who worked as a household servant and spoke a third language that the visitor understood. This unnamed Portuguese-speaking woman then described to the court what she had learned from the new arrival, including a story about unicorns—referring to rhinoceroses, or covertly mocking gullible Lusitanian listeners?

Once able to understand the new arrival's speech through a third language, other African interpreters began to learn his language and teach him basic Portuguese vocabulary and grammar. After these language exchanges, the unnamed man from the Mesurado peninsula learned sufficient Portuguese for the next step. A delegation escorted him on a tour of the kingdom to dazzle him with the benefits his compatriots might obtain through trade. Having sufficiently impressed the captive, the Portuguese returned him to his home on a different caravel. They intended for him to serve as a translator thereafter and, presumably, convince his compatriots of the advantages of engaging in trade with the Portuguese.[28]

Seizing individuals to serve as interpreters and as commercial liaisons soon became standard protocol. After decades of failure under Prince Henry, King Afonso had finally found the formula for commercial success. Over the next forty years, Portuguese commanders enticed or seized countless speakers of unfamiliar languages from the coasts of Africa during initial contact. These individuals were taken to Lisbon, where locals would learn some of their vocabulary while teaching them Portuguese, which was suitable for commerce. Hence, they were treated very differently than individuals captured for slavery. The visitors would be shown various goods that the Portuguese had produced or acquired. Their escorts paid careful attention to the goods the out-of-towners admired and likely gave them samples to take back to their home communities.[29]

Thus, the Portuguese abandoned efforts to seize goods and people by themselves; instead, they empowered native leaders to act as their intermediaries in acquiring enslaved people and goods. The primary objective of initial contact shifted from merely exchanging trade goods (as Cadamosto had done along the Gambia) to finding friendly coastal communities who possessed such goods in commercial quantities and, just as significantly, had a leadership or social structure capable of and amenable to supplying visiting Portuguese ships.

Knowledge of Indigenous languages and cultures became the linchpin of Portuguese commercial success overseas for the remainder of the fifteenth century.[30] From the beginning, infrequently, some Portuguese subjects voluntarily went to live among the natives in the hope of gathering sufficient information about trade and other practices to bring back to the Crown. What rewards they might have expected to receive upon their return are not specified in any surviving text.[31]

LANGUAGE: THE KEY TO EMPIRE

In 1492, the Spanish humanist and grammarian Antonio de Nebrija wrote, "Language is the companion of empire."[32] When he composed this epigram, quoted many times since, Spain did not yet have dominion over scores of peoples speaking dozens of unfamiliar languages—their foreign subjects all spoke the long-familiar Arabic. However, by the time he composed this pithy formula, Nebrija had observed for decades his Portuguese neighbors' overseas economic success, and how they had taught their language to countless communities abroad, while learning valuable Indigenous idioms as well.

Following Sintra's return, the pace of voyages appears to have slowed. Most explanations for this downturn invoke the death of Prince Henry, but a more profound reason resides with the dramatic changes in the ocean itself. Where Portuguese ships had previously traveled, the winds blew predictably: "From Cape Bojador to the Isles de Los, the winds . . . follow the direction of the coast from North to South, and . . . seldom vary."[33] But south of Cape Verga, sea breezes presented an entirely different challenge. Sintra had managed to dodge these hazards, but sailors following his route had to contend with a new, often lethal obstacle presented by either the wind or its absence, a phenomenon later known as the doldrums.

6

A Painted Ship upon a Painted Ocean

1460–1470

Describing what sailors experienced when nearing the equator, Samuel Taylor Coleridge wrote:

> Down dropt the breeze, the sails dropt down . . .
> Nor breath nor motion; [the ship]
> As idle as a painted ship
> Upon a painted ocean.

Sailors had good reason to dread this stillness. Humans can survive only three, perhaps four, days without water. The longer their ships remained motionless, the more deadly the threat of dehydration became. Two famous lines capture that rising panic:

> Water, water, everywhere,
> Nor any drop to drink.[1]

These areas would later be christened "the doldrums" for the misery and depression that sailors felt. Yet they could do nothing to evade this natural occurrence.

Around the middle of the planet, winds from the north that spin clockwise collide with winds from the south that spin counterclockwise. When these winds meet, moisture-filled air rises from the ocean, creating very little wind for sailors. High in the sky, this leads to constant thunderstorms. Scientists call this area where winds meet over tropical oceans the Intertropical Convergence Zone.[2] Even though these winds clash near the equator in all oceans, they threaten navigation in the Atlantic more than in any other ocean.

Opposing winds prove favorable on several seas other than the Atlantic, promoting annual monsoonal reversal of winds and currents, thus bolstering seagoing trade in the Arabian and South China Seas for millennia.[3] However, no large-scale monsoons appear in the Atlantic.[4] In the Western Pacific, monsoons were similarly absent, but the still winds and thunderstorms near the equator sporadically interfered with oceangoing Polynesians. Their double canoes confronted a steady eastward-flowing current that allowed travel from the Society Islands to Hawaii and back.[5] Yet neither option emerged in the Atlantic, where a dead zone twice the size of the Polynesians' confronted would-be sailors and where the helpful eastward equatorial currents lay out of reach, hundreds of miles to the south.

Even today, these winds imperil Atlantic navigation—but instead of sailing vessels, they threaten modern jets. The thunderstorms around the equator tower, often reaching fifty-five thousand feet—ten thousand feet higher than most commercial airplanes can fly. Most planes avoid these high-altitude storms, but in June 2009, an Air France jet flying over the mid-Atlantic ventured into one. When the plane's speed indicators froze in the storm, the pilots did not know how fast they traveled. Believing they were flying too quickly, they mistakenly slowed the plane until it was no longer moving forward. The giant plane fell out of the sky, killing all 228 on board. As it slammed onto the Atlantic, raindrops splashed in tiny circles, but the surface waves remained calm.[6]

In the same seas where Air France Flight 447 crashed, fifteenth-century sailors first encountered the deadly doldrums. Initially, they would have been bewildered. In the Mediterranean and the northern reaches of the Atlantic, where Europeans had previously navigated, breezes returned swiftly after occasional lulls. But here, the first boats trapped in the doldrums would have drifted aimlessly until their sailors perished from thirst, leaving only skeletons on a ghost ship silhouetted against a dark, glassy sea—boats, captains, and crews forever lost. By 1460, Portuguese sailors had discovered a seasonal and geographic pattern to these deadly calms, and by finding a way to navigate through the doldrums that plagued the Atlantic, they also located a route to a unique African shrub that produced the first commercially valuable botanical export.[7]

Seen from space, these high-level storms resemble puffs of white smoke around the middle of the planet. If we think of the atmosphere as a person breathing, she inhales deeply from the warm water of the

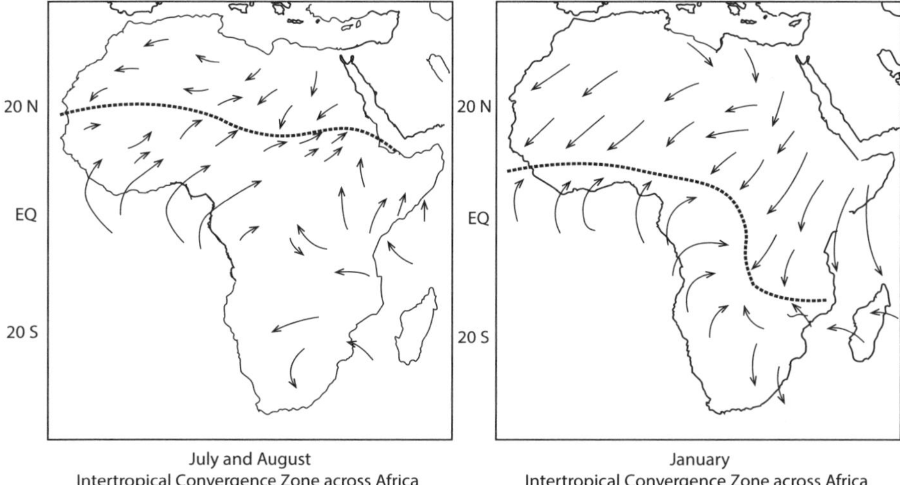

MAP 7. The Intertropical Convergence Zone across Africa in winter and summer. After Nicholson, *Dryland Climatology*, 2011.

ocean before exhaling it into massive, fluffy clouds above. As the atmosphere inhales, winds blow not north, south, east, or west, but rather straight up, leaving the surface still. When the exhaled puffs fill the atmosphere, they dump their moisture into the ocean during thunderstorms. As the rains pour, the winds blow erratically but in no specific direction.[8]

As these clashing winds drift over terrain, they encounter countless natural obstacles: elevation changes, mountain ranges, valleys, bluffs, and cliffs. All redirecting the winds, these obstacles divert them from some places and channel them elsewhere. The most dramatic land diversions occur over Africa. During the Northern Hemisphere's summer, the winds clash over the Sahara, more than fifteen hundred land miles (twenty-five hundred kilometers) north of the geographic equator. During winter, this still zone moves south but never even reaches the equator in western Africa, remaining around 750 land miles (twelve hundred kilometers) to its north.[9]

However, on the continent's eastern side during northern winter, this belt of colliding winds dives steeply southward more than eleven hundred land miles (nineteen hundred kilometers) south of the geographic equator, nearly to the latitude of Antananarivo, the capital of Madagascar, 4.5 degrees (271 nautical miles) past the southernmost boundary of the classical torrid zone.[10] Hence, the still weather and seasonally pouring rains over a land area nearly three thousand miles north and south of the equator. However, these dramatic shifts are not transferred to the Atlantic Ocean, exactly because land and ocean absorb heat at different rates.[11]

THE EQUATOR

Most of us associate Earth's hottest temperatures with the geographic equator. However, like Aristotle's original concept, this belief links the heat with a fixed place on Earth that classical scholars called the torrid zone, popularly believed to lie on the equator.[12] However, the sun's constantly changing position throughout the year shifts this zone to the north and south. Seen from Earth, the sun appears to pass back and forth continually between northern and southern boundaries, but Earth's annual route around the sun is responsible for these daily changes.

The sun's direct overhead rays arrive at their most northerly position on Earth, at the Tropic of Cancer, at a latitude of around 23.5 degrees. At this moment, the Northern Hemisphere is tilted maximally toward the sun at a day and time astronomers call the June solstice. In the other hemisphere, direct rays reach their most southerly position of −23.5 degrees, known by navigators as the Tropic of Capricorn, on a day astronomers call the December solstice.[13] Thus, the Tropics of Cancer and Capricorn define the boundaries of zones of maximum heat.

Early in their southward voyages, mariners realized that the zone of potential stillness lay within these boundaries. In 1495, a visitor to Portugal reported that "after crossing the Tropic of Cancer [23.5 degrees north], ships will always encounter motionless seas with little wind," adding simply that "the reason for this is unknown."[14] Because the sun's rays are most intense when directly overhead, it seems reasonable to assume that the air and water would become hottest on those days; unfortunately, the sun's position overhead rarely corresponds directly with the hottest temperatures on Earth.[15]

THE HEAT EQUATOR

For over a hundred years, scientists have added a modifier to the word *equator* to designate the continually moving zone of scorching temperatures connected to the sun's position overhead.[16] Since the 1940s, nearly all have adopted the words *heat equator* or *thermal equator* to refer to those places where average air temperatures are hottest on any given day. However, rather than a neat circle like a hoop sliding up and down a pole, this heat pathway loops up and down across both landmasses and oceans.

Anyone who has ever sat waiting for a pot to boil knows how frustratingly slow it can seem for the water to heat up. Oceans are much larger versions of that pot that takes forever to boil. They absorb heat far more slowly and, fortunately for us, never reach temperatures as hot as those on land. Thus, oceans absorb much of the heat generated in the atmosphere without raising ocean temperatures to the same degree. In this way, oceans are crucial to life on Earth, keeping the planet's temperatures cool. The next time you become exasperated waiting for a kettle to boil, you might think how that phenomenon is essential to what makes life habitable on Earth.[17]

In both hemispheres, the ocean needs at least thirty days past the solstice to soak up the sun's heat thoroughly. Thus, in the Northern Hemisphere, Atlantic Ocean temperatures soar in July and August (after the June solstice); whereas in the Southern Hemisphere, they surge in January and February. Although reaching farther north than such zones in any other ocean, the doldrums rarely went beyond Cape Verga, ten degrees north of the equator. During summer, they continued southward for six hundred miles until they barely skimmed the equator.[18] As winter arrived between November and April, warmer waters began retreating slowly until February, when they covered less than a hundred miles. If a ship waited until the winter months, it could limit its time in the doldrums. Such a short interval could not be guaranteed, though; in some years, anomalies in temperature, dust clouds, and surface evaporation caused winds to shift unexpectedly. Still, for the most part, the Atlantic doldrums were best approached during winter.

During the 1460s, Portuguese sailors developed a strategy for surviving the doldrums, given that, at unexpected moments, any ship might run into an unexpected zone of still winds.[19] Their strategy relied on remaining a safe distance from shore (no more than two hundred miles) but taking advantage of the opposite directions of the winds during the day and the night.[20]

As the sun rises, land heats up faster than the ocean, creating an area of low pressure. Breezes blow from the chilly ocean to the warmer land during the day. Conversely, when the sun drops below the horizon, land cools faster than the ocean, which also warms during the day, reversing the direction of the breezes. "Vessels sail by day with little wind and at night without any," a visitor declared in 1495, adding, "It is necessary to anchor at night when there is no wind." However, from Cape Mesurado onward, countless boulders lay hidden just below the water's surface. Even with a night breeze, night sailing proved impossible, as sailors risked slamming into deadly hidden obstacles. Hence, sailors obeyed advice to remain stationary even when nighttime breezes occasionally arose.[21]

The first expedition to put Afonso's new emphasis on establishing trade and understanding Indigenous languages into practice was also the first to pass through the doldrums unscathed. In 1460, Pedro de Sintra managed to dodge the calms entirely, which means that he likely was sailing between late fall and early spring. Had he traveled at the height of spring or summer, he would have first encountered doldrums near the diamond-shaped Cape Verga, at ten degrees north, and struggled against them until reaching the so-called Woods of Saint Mary, where he made a historic reversal and returned to Portugal.

Although Sintra would not have known it then, he could have traveled farther between November and April. In February, a doldrum-free zone continued for another 140 miles beyond where he had turned around, as far as the Sinoe River in present-day Liberia.[22] Despite Sintra's success in timing his voyage southward, new commercial opportunities eluded him. That prospect emerged shortly after, with opportunities just north of the doldrums. Soon after his return, other vessels negotiated the coast south of Sintra's final stopping point at the Woods of Saint Mary. At first, they saw nineteen miles of "sandy beach backed by thick brush and trees," terminating in a delta where three rivers met. Portuguese maps labeled this confluence as one, the Junco, later anglicized as the Junk River.[23]

Trees surrounded the Junk River entrance, unlike the mangrove-shrouded mouths of rivers to the north. On the west bank lay an excellent cove; on the other, several smaller villages spread out, including one with exceptionally fresh water. Despite the tranquil harbor, crews traded only for small quantities of gold and a pepper-like spice at the entrance to the Junk.[24] South of the river, countless boulders lurked just below the surface. Therefore, guidebooks recommended that pilots approach land only during daytime if they sighted native canoes, an indicator of safe passage through the waters for the rowboats they carried.[25]

Only a few pepper-like spice plants could be found at the Junk River, but soon the Portuguese would find a new supply. This two-foot plant with trumpet-shaped purple flowers grew in large quantities south of the Junk. After blooming, the flowers turn into small pods with scores of tiny reddish-brown seeds. Although the plant belongs genetically to the ginger family, the little seeds taste like pepper with a citrusy flavor. Sometimes known as "grains of Paradise," this spice found a market in Europe for several centuries, where it served as a cheaper substitute for the vastly more expensive Indian black pepper. As a result, much of the area became known in Portuguese as the "Grain [of Paradise]" or "Malagueta [Pepper]" Coast.[26]

Just south of the Junk resided a highly organized native community, the Grebo (also called Reguebos or Zeguebos), who were willing and able to harvest and exchange vast amounts of this spice.[27] Their sophisticated operation contributed substantially to the booming trade in this fast-growing wild plant. When the large Portuguese vessels reached the mouth of the river where the Grebo resided, sailors encountered dirty, muddy waters that obscured the seafloor. Even though Portuguese ships needed to anchor at a distance from where the Grebos resided, the Europeans' shelter was secure and relatively free of obstacles. So, at the J-shaped entrance to the river, the Portuguese ships sent smaller boats one-and-a-quarter miles upriver to meet the canoes and trade for the baskets of plants.

Grebos met the Portuguese boats with a fleet of canoes, each brimming with baskets containing packages of the red seed pods carefully wrapped in their leaves and neatly tied into squares using the plant's stems, each packet weighing between ten and thirty pounds.[28] Their long, narrow, pointed canoes each held three or four men who rowed while kneeling. One observer noted that while many natives elsewhere were hesitant to trade where "Malagueta [grew,] these natives travel to the ships."[29] Grebos approached and often boarded the Portuguese boats, exchanging their pepper-filled bushels for goods the Portuguese carried. These pepper-seed containers became so familiar that the Portuguese named the river where they traded River of the Baskets (Rio dos Cestos). To this day, the river's name remains a variant of *Cestos:* the River Cess.[30]

Two other coastal places also supplied considerable quantities of the spice. A Genoese man with a Portuguese name, Soeiro da Costa, found immense amounts of Malagueta pepper growing wild in swampy areas near the coast, where he began trading.[31] That location became known as Soeiro's Trading Post, or occasionally "the Genoese Trading Post."[32] Farther south, just past the Sinoe River, the northern edge of the seasonal doldrums, a second community also had access to large quantities of Malagueta seeds. Again, because of the dangerously rock-strewn coast, Portuguese ships had to drop anchor at a considerable distance from shore. The men of this coastal town fished in deeper offshore waters and

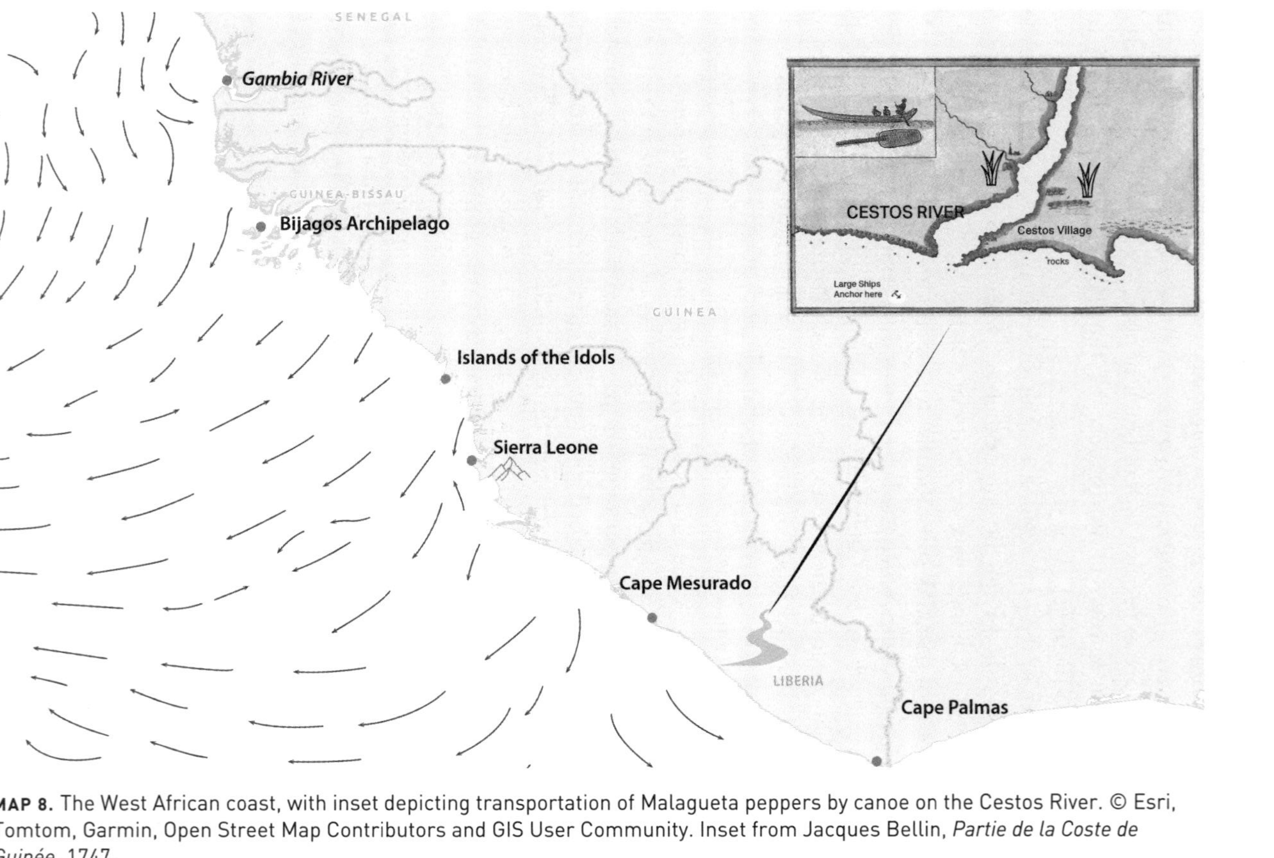

MAP 8. The West African coast, with inset depicting transportation of Malagueta peppers by canoe on the Cestos River. © Esri, Tomtom, Garmin, Open Street Map Contributors and GIS User Community. Inset from Jacques Bellin, *Partie de la Coste de Guinée*, 1747.

were adept at navigating difficult seas in large, purpose-built canoes. Thus, unlike the pepper sellers at the River of the Baskets, the Kru fishermen here could row out to meet the larger vessels in their deep-sea fishing canoes, where they traded citrusy pepper with Portuguese merchants.[33] Their skills would later lead them to their recruitment as sailors on British ships, where they developed a reputation for talented seamanship.

When black pepper from India became more widely available, the Malagueta seed declined in popularity. However, the plant left its mark on the names of many places, appearing on maps for centuries after the commerce had withered away as the "Grain Coast" or "Malagueta Coast."[34]

In addition to the pepper trade, a small but significant slave trade also took place off these shores. Immediately following the Cestos River came a series of six rocks and reefs, at the end of which was a tiny islet called Palma, now Baiyan, where fishermen sold enslaved people belonging to a community they called the Guey.[35] They held the Guey in a small village on a tiny river named Bioni and transported them from the shore to the islet in their boats while the larger slave ships remained in deeper water.[36] Another slave market lay close by. Past the Kru fishermen came a broad stretch of sand called the Beach of Slaves. During the first decade or two of contact with the Portuguese, large-scale slaving occurred along this beach. In the middle of this lengthy shore came a River of Slaves (now the Dugbe) that reached the Atlantic and was likely the origin of these enslaved people, about whom little is known. By 1475, this slave trade along the shore had virtually ended, but the designation "Slave Beach," like "Grain/Malagueta Coast," remained on maps for centuries.[37]

A humorous incident in the mid-1470s demonstrated how well the trading partners on this coast understood their visitors. They recounted to a visiting Portuguese crew how a competitor's boat, laden with gold stolen from Mina, became becalmed, snapped its anchor on the rocks, and drifted helplessly toward the beach. The locals then described how they had eaten the thirty-five Flemish crew members and their Spanish pilot. Delighted Portuguese listeners asked them multiple times to retell this possibly apocryphal story of their enemies being eaten alive, which they did. True or not, the story showed the cultural significance of revenge cannibalism to the storytellers' rapt listeners.[38]

However, the environment restricted trade in this area. If a ship was short on firewood and water, a small river behind Bafu Point could supply both.[39] But ultimately, half a century later, the great Portuguese captain and cartographer Duarte Pacheco Pereira warned ships to avoid this stretch of coast entirely between May and the end of September because of the misery caused by thunderstorms—a sign of the doldrums.[40]

The challenges of getting close to this coast forced Portuguese ships to anchor far from shore. Members of coastal communities who lived near rivers or on accessible islands, or who had good offshore navigation skills and large canoes, were thus expected to transport people or spices to the ships. Additionally, these communities needed to be well organized if they were to manage the gathering and moving of people or pepper. Many other coastal communities, lacking big canoes, organized systems, or valuable goods, stayed beyond the reach of European trade networks.

During the 1460s, licensed Portuguese ships had located nearly every source of pepper plants and soon-to-be enslaved people along the coast. Unsatisfied, by the decade's end, King Afonso authorized a Portuguese merchant named Fernão Gomes to lead ships into the next challenging section of the coast. This voyage would entail rounding the infamous Cape Palmas, whose 2.4-mile-long stone tail blocked their route. But as the ships he sponsored looped around the protruding rocks, navigators confronted another new challenge: the Atlantic's intricate equatorial currents. One such swift-moving current would carry them into the heart of Africa's legendary gold fields and take the slave trade to new depths of exploitation.

7

Gold at Last

The Route to Mina, 1470–1480

Dutch trader-explorer Pieter de Marees's centuries-old caution that gold mining "costs many people's lives" remains true today. Fatalities still occur in small-scale mines like those that operated historically throughout the gold belt—along significant rivers, in tunnels, and in excavated shafts in remote vegetated areas. In one incident that made international news, a tunnel collapsed in western Ghana, killing half of the thirty people working in the gold mine; all but two of the fatalities were women, employed to carry gold from the tunnels and shafts.[1] In the fifteenth century, discovery of harbors close to the gold mines came unexpectedly, shortly after Portuguese pilots entered the Gulf of Guinea.

HURTLING ALONG THE GUINEA CURRENT

Thirty-five years after the first Portuguese ship rounded Bojador, boats venturing south along the African coast were suddenly careening in a new direction, a change that surprised them. While anyone today who has seen a classroom map knows that the west coast of Africa curves to the east, in 1470 it baffled the first crews who were suddenly trapped by a swift eastbound current. As a navigator explained in 1507, "Sailing onward from the coast of Cabo das Palmas [differs] because from this cape you sail in one direction different from the one along the Malagueta [Pepper] Coast."[2] As he noted, this unexpected turn occurred at the "Cape of Palms," so named by Diogo Gomes and today known as Cape

Palmas. The anodyne name conceals a treacherous reality: The cape was protected by a perilous string of giant boulders that spelled danger to any passing ship. Attempting to halt near the rocks would have risked losing an anchor, thus endangering the entire boat and crew.

From the earliest account in 1507 through to the latest scientific survey, all observers have reached the same conclusion: The Guinea Current is formidable at the cape, its speed peaking at thirty-six to forty-two miles per hour daily. Historical reports mention that the current "flowed fastest in the vicinity of Cape Palmas . . . reaching 85 miles per day [6.2 knots] during June," with speeds varying daily in other months.[3]

These fast waters made approaching the primary source of gold difficult. The earliest instructions cautioned pilots that, if they wished to find the harbor for gold trade, they would have to battle the momentum of a current that seemed hell-bent on shoving them east. A 1507 guide warned: "Two to three leagues away [from your destination], there are two red cliffs. . . . Without having previously sailed there, look for high hills on the east side that extend to Rio da Volta. And you will see these signs when you are *past Mina*," the principal gold-trading post.[4] But speed was not the only obstacle. During October and November, the current stopped at the Cape of Three Points, circled 180 degrees, and turned back to Cape Palmas, making it impossible to sail farther. At the start of December, the current resumed its customary eastbound speed until the following October, when the cycle repeated.[5]

Nearing the eastern end of the Gulf of Guinea, the current split in two: One part headed north to the coast near the polity of Benin (eastern Ghana, Togo, Benin, and western Nigeria), while the other section continued straight across the gulf, almost to the coast of Cameroon. Thus, when King Afonso V gave the go-ahead to explore what turned out to be the Gulf of Guinea, progress across the gulf was swift. In 1469, a Portuguese ship reached the tiny island (modern Bioko) only seventeen nautical miles (thirty-two kilometers) from the west coast of Africa. Its towering peak, nearly ten thousand feet (three thousand meters) high, made it easy to spot from miles away. But most startling to anyone looking at a map is the distance to modern Cameroon: The eastbound current was such that in a little over two weeks after rounding Cape Palmas, a ship could anchor on the coast in the Gulf of Biafra.

Looking at any map, this early landing over two hundred miles away may seem puzzling, given that it bypassed many miles of coast and countless excellent harbors. However, the ocean can vex the unwary map reader, with currents determining which lands should be reached

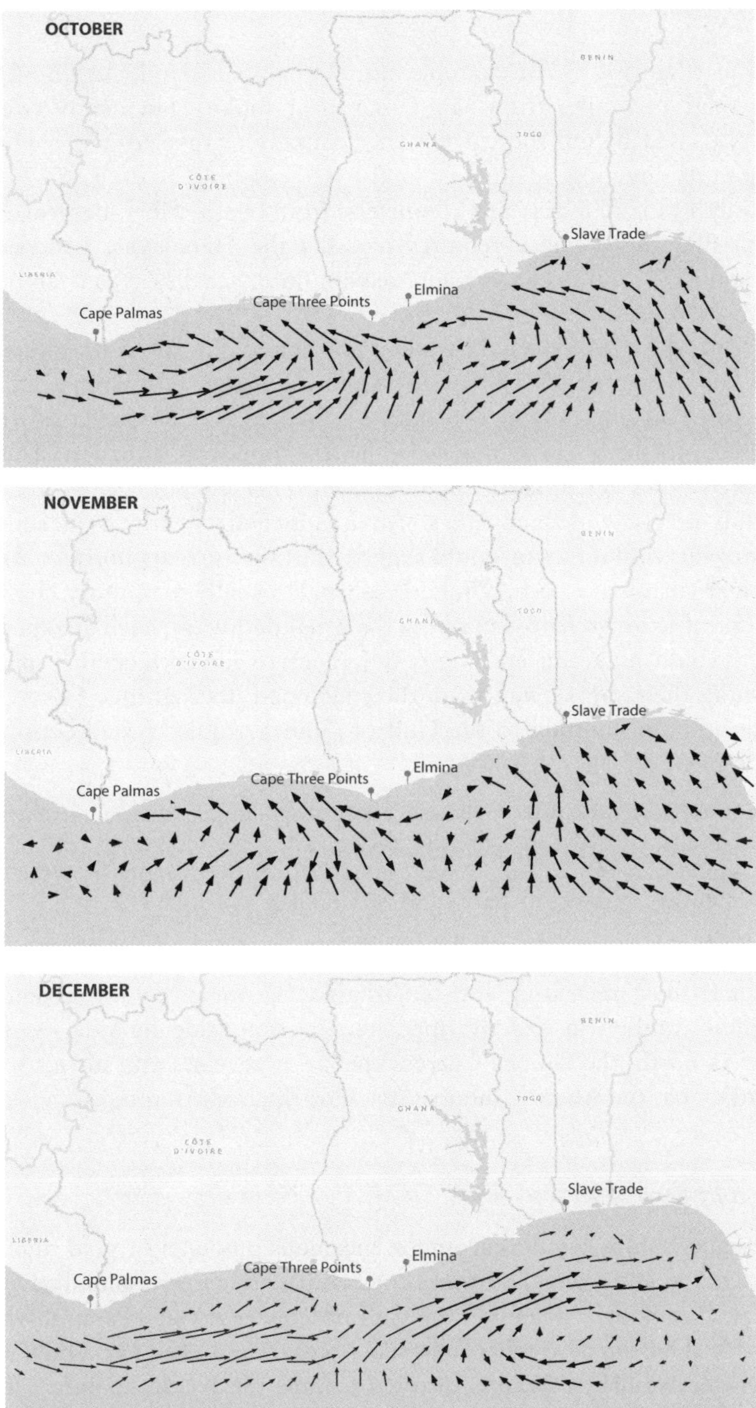

MAP 9a–c. The Guinea Current, October–December. © Esri, Tomtom, Garmin, Open Street Map Contributors and GIS User Community. Data from Bosson et al., "Surface Guinea Current Variability," 2023.

and in what order. For example, an uninhabited island (Ano Bon) lay less than 110 miles from São Tomé (near Bioko), but nearly twenty years passed before the currents accidentally channeled a Portuguese ship in its direction.[6] To take a land-based example, Bhutan and Nepal lie only 153 land miles (247 kilometers) from Lhasa, Tibet, but traveling from Bhutan to Lhasa requires crossing the Himalayas. Landforms remain rigid, and ocean currents remain fluid, yet they create similarly formidable obstacles.[7]

While the trip to the gold-trading post posed difficulties for pilots, so did the voyage back. The speedy Guinea Current was comparatively narrow, reaching only as far south as three degrees north. At just two degrees north, a current traveled in the opposite direction. Unlike marked lanes on modern highways, currents do not always remain within precisely defined lanes. An unanticipated eddy or a sudden change in wind direction could slightly shift the currents north or south of their customary path. While these small, usually temporary changes matter little to a scientist marking the usual pathways, such unexpected shifts could make an enormous difference to fifteenth-century pilots, pushing their ships away from their intended destinations. Hence, on the outbound journey to the Gulf of Guinea, guides issued continual warnings

> not [to] descend further to the south than 2° N. latitude, so as not to encounter the equatorial current setting to the W. . . . As soon as the parallel of Cape Palmas is reached, . . . you can navigate at any distance from the land but always to the northward of 2° N. latitude.[8]

The space separating the Guinea and Equatorial Currents is generally limited, thus presenting the remarkable feature of two well-defined streams running in exactly opposite directions side by side. At two degrees north, the Guinea Current speedway reverses and travels westward across the Atlantic (almost reaching the mid-Atlantic).

GOLD AT LAST

Portuguese ships would encounter enormous amounts of gold about a third of the way along the Guinea coast. But getting to it required navigating the headlong plunge into the Gulf of Guinea at Cape Palmas, where the first three hundred miles followed a "coastline arched like a bow with three arrows (Three Points) directed toward the South Atlantic." Two giant whirlpools periodically formed within this arched bow, slowing eastward voyages. The trio of rocks extending into the water, aptly named the Cape of Three Points, provided the first notable landmark and was the only cape extending any significant distance into the gulf.[9] Just east of this cape lay a bay (three and a half miles wide) that provided the best anchorage between Sierra Leone and the islands on the opposite side of the gulf. Unfortunately for the Portuguese, while this area offered shelter, it presented no new commercial opportunities in the vicinity.[10]

Eventually, sometime between 1471 and 1474, a Portuguese expedition succeeded in sailing close to the shore, navigating the reefs and rocks surrounding the third arrow of the Cape of Three Points. Thirty miles east, they anchored off one of many unremarkable peninsulas, beside a small inlet. During the eleventh century, a diverse group of Akan speakers had settled throughout the rich gold-bearing lands of southern Côte d'Ivoire, Ghana, and Togo.[11] Some occupied small independent polities, while others (the Ashanti and Dinkira) dominated several other gold-rich realms.[12] When the Portuguese first arrived, on opposite sides of this short stream sat two sets of houses inhabited by Akan people with political allegiance to rulers of different communities, one to the leader of the nearby Eguafo and the other to the Fetu farther east.[13]

But nothing about this place's natural or built environment, at least when seen from a distance, prepared the Portuguese for what they would find on shore. Following their usual practice, they sent a small contingent of men in advance to determine whether the locals would welcome their presence. To their shock, curious natives from both villages appeared, wearing large quantities of high-quality gold jewelry and decorations.[14] The Portuguese visitors had stumbled upon a pair of coastal villages with a long-established route to West Africa's legendary massive gold fields.[15]

The Portuguese soon realized that no gold mines existed on the peninsula or in the towns. Instead, the gold came from the same two-billion-year-old geological formations that had built the settlement at Djenne, enriched Mansa Musa, the legendary thirteenth-century lord of Mali, and furnished the gold that caravans moved north across the Sahara to ports along the Mediterranean during the Middle Ages. Since the waters by this peninsula lay far closer to the gold fields than any land route to Europe, this discovery promised a far faster trade route than the arduous overland trips through desert or jungle. Portuguese historians have called this revelation the "triumph of the caravel over the caravan."[16]

Various forces conspired to make the Twin Villages, as the Portuguese initially called this area, a hotspot for the gold trade.[17] In addition to safe anchorages easily reached by local canoes, it had the fastest and safest route to the mines by far.[18]

When the first ship laden with gold returned to Portugal early in the 1470s, authorities found that it was impossible to keep the news of this discovery secret. Around the world at this time, many countries were experiencing a shortage of gold for coins, causing them to turn to copper (in Egypt) or heavily restrictive government regulation (in England).[19] When Portuguese merchants suddenly began presenting quantities of high-quality gold for purchases abroad, other Europeans wanted to know the source. And soon enough, many of Europe's merchants and aristocrats became aware of the location of this new font of gold and began to compete with the Portuguese for access to it.[20]

The stakes of this discovery were enormous, promising to reshape the world's economy. Until well into the nineteenth century, what would become the town of Elmina, on the south coast of Ghana, would supply much of the world's gold and most of Europe's, only briefly eclipsed by soon-exhausted Caribbean discoveries (sold to the Portuguese in the Azores, placing them in control of the Atlantic gold trade). Even today, the quantity of legally exported gold makes the geographically small area of Ghana the world's seventh largest gold exporter, just behind such large countries as China and Australia. Other powers wanted their share of the discovery. Thus, as news of this sudden new source of gold percolated through an area precariously short of the mineral, other Europeans attempted to bypass the Portuguese and travel directly to the source. Records of two such attempts in the 1470s survive, one by the Castilian Crown, the other by Flemish merchants.

Most voyages to Mina were privately funded, but a significant exception was the trip sponsored by Queen Isabel of Spain. To understand why she chose to fund this voyage and how it reshaped Portuguese policy, we need to delve into the complex dynastic politics of fifteenth-century Iberia.

In 1468, King Henry IV of Castile, who reigned from 1454 until his death in 1474, faced a difficult decision. He had a two-year-old daughter named Juana, the legal heir to the crowns of Castile and Leon. Given her very young age and the high likelihood that she might not survive to adulthood, Henry promised his throne to his twenty-three-year-old half-sister, Isabel, on the condition that she marry someone of his choosing. Deceitfully, Isabel agreed, only to secretly wed Ferdinand II, heir to the eastern kingdom of Aragon and a veteran of multiple wars. King Henry responded by disinheriting her and naming his daughter Juana heir to the throne of Castile. However, the Castilian nobility recognized in Isabel the backbone of steel they wanted in a leader. Hence, when King Afonso V married Juana, the sole daughter of Henry IV, and invaded Castile in May 1475, the combination of Castilian and Aragonese forces backed Isabel, leading to a stalemate for the throne of Castile. Following a series of indecisive battles, Afonso abandoned his overland advance.[21]

Within days of Afonso's incursion, Isabel of Castile responded by launching a spurious claim to the gold fields of the Guinea coast. Knowing that the Portuguese had failed to fortify the site, Isabel spuriously claimed that her Lusitanian rivals had for centuries "interfered with" her rightful "conquest [of gold]" and that King Afonso was taking a 20 percent tariff "to the detriment of . . . [her] income." She proposed to "separate this conquest from the said adversary, Portugal, and make war . . . with all the evil and damage as [befits] an adversary . . . and to apply the fifth [i.e., Afonso's tariff] to my income." Isabel sent a Castilian fleet of thirty-five ships, including merchant vessels, to enforce this claim. They arrived at the principal port without encountering Portuguese resistance and collected large amounts of gold. Indeed, the potential revenue from the Guinea coast was so significant, and her expectations of victory so high, that she appointed two men to collect the 20 percent tariff in Seville even before the attack.[22]

Responding swiftly, Afonso organized eleven ships to intercept the Castilian expedition before it could return to Spain. Under the command of knights Jorge Correia and Mem Palha, the Portuguese fleet, despite being outnumbered, demonstrated their naval expertise and firepower, easily defeating the more extensive Spanish fleet at Mina. This victory, along with the successful defense of their Atlantic island bases, showcases the resilience and determination of the Portuguese in the face of adversity.[23]

After their land battles ended in stalemate and the naval victories went to the Portuguese, the two sides signed a peace treaty whereby Isabel withdrew her bogus claim to the Guinean gold fields. At the same time, Portugal retained control of all locations accessible only by sea (except the Canaries). Isabel retained control of her boundaries on the peninsula, which more or less conform to the modern borders of Iberia. This ambiguous conclusion, with Spanish victory on the ground and Portuguese triumph on the water, foreshadowed the future differences in the respective empires. Spain achieved military successes on land in the Americas, while Portugal established far-flung coastal territories from the Atlantic to the Indian Ocean, accessible only by sea.[24] Ironically, just forty-one years after this war, in 1516, John III, who was both Isabel's great-grandson and Afonso's great-nephew, ascended the throne of Castile, thus ending even the remotest chance of reviving the dispute.[25]

The conflict with Spain prompted the Portuguese officials to recognize the vulnerability of their trade. In response, the Portuguese Crown took decisive action, ordering the construction of a coastal fort to protect their Gold Coast holdings from large-scale European incursions and to safeguard gold collected from the interior. This strategic fortification, completed in 1482, was known as "São Jorge da Mina," or simply "Mina." While it effectively deterred large-scale attacks for over a century, it was less successful in discouraging piracy, a lucrative venture for pirates from seafaring countries along the Atlantic coast. One of those failed freelance expeditions produced the only surviving account of how the Portuguese acquired gold at the wealthiest port along the coast. In 1480, a Flemish adventurer captured at Mina was transported to a prison in Portugal. After escaping, Eustache de La Fosse sought refuge in France, where he informed his hosts of the Portuguese methods of gold trading he had observed, which had unexpected implications for the slave trade.[26]

This episode began on the Sunday before Christmas 1479, when three Flemish ships were anchored near the principal gold-trading port. Some of their crew had disembarked, carrying goods they intended to trade for gold. However, the crewmen had to wait offshore for a week because their visit was unexpected; news of the ships' arrival had to reach the gold miners in the interior, who then had to dispatch their representatives to the coast. After traders began arriving, the Flemish exchanged their goods for gold during the next two weeks. At the fortnight's end, two ships had as much as they could carry and had already departed when misfortune struck. On January 6, 1480, four heavily armed Portuguese ships seized both vessels, whose crews could offer little resistance.[27]

The Portuguese released the ordinary seamen and poorer crew members from the Flemish fleet, placing them aboard a ship with one sail, one anchor, some water, and hard bread—effectively an execution. Surprisingly, at least one person on this ship managed to navigate exceptionally well, because the vessel returned to Spain despite having only a single sail and anchor.[28] Having dispatched the people they viewed as redundant, the Portuguese placed their ten most valuable prisoners, including La Fosse, aboard a ship commanded by the soon-to-be-famous Diogo Cão, which remained offshore while two other ships were sent to search the coast for enslaved people to bring back to Mina to mine for gold. Here, La Fosse's writing gives us a rare insight into the trades in gold and human beings in West Africa.

The speedy path of the Guinea Current determined their destination. Although numerous harbors and communities along the way would have been able to furnish the additional labor required by the increased demand for gold, the swift waters quickly sent these vessels to a place where the current moved close to the coast, past the reversing waters of the Rio da Volta and near the empire of Benin, later a significant center of the slave trade. In February 1480, two Portuguese ships rapidly acquired four hundred enslaved people.[29]

After dividing the four hundred people evenly between the two ships, the caravels returned to Mina six weeks after their arrival. Upon their return, they sold most of the enslaved people to the native owners of the gold mines at double what they had paid to acquire them. In addition, they sold a smaller number of enslaved people, including women and children, captured from the Flemish fleet.[30]

The Flemish seamen had purchased enslaved people in advance of their arrival at Mina, having stopped along the Pepper Coast somewhere south of the River of the Baskets (Cestos), where one sailor boasted of paying residents only a barber's basin and three or four large brass rings to buy a mother and child. Once at Mina, he claimed that he sold the mother and child for twelve or fourteen gold weights each, which suggests he knew that native miners purchased women and children to labor in gold mining. While no fifteenth-century descriptions of gendered labor in gold mines survive, such specifics appear in the first Dutch reports on the roles of enslaved men and women at Mina.[31]

In Ghana, centuries of weathering had crumbled the gold-bearing rock close to the surface. Buffeted by wind and rain, fragments of the precious metal had splintered off and tumbled into rivers. There, women sifted out the stones to search for slivers of gold. The rivers carried great amounts of the valuable fragments, which heavy rains swept out to sea from April to July; beginning in August, women had the laborious task of sifting through beach sand at low tide for traces of the gold deposited by the rushing waters. In 1668, the Dutch writer Olfert Dapper wrote: "Gold is deposited on the beach and gathered at low tide by women who sometimes find pieces worth two and three guilders. While their labor yielded only small amounts, larger fragments of weathered gold remained in the rivers where men were sent to retrieve them."[32]

Enslaved men and women collected this alluvial gold, but some men were sent on more dangerous retrievals. Throughout the gold-bearing regions, water torrents cascaded over the hilly landscapes. That force fractured the gold-yielding stone into smaller pieces that sank to the bottom of the whirlpools that formed at the base of cliffs. Dapper continued:

> The black [men] bring it to the surface. They dive with a wooden trough after all the earth and stones they can find, then come up with the trough on their heads and exhale. Then they wash [the rocks] out to see the gold, the heaviest material, at the bottom of the trough. From time to time, the violence of the pouring water washes off whole pieces from the exposed gold-running veins. Sometimes, they find many gold pieces while diving, but sometimes, they dive all day in vain.[33]

Finally, a third and more dangerous method involved hacking into the stone where the gold lay. In 1602, Pieter de Marees wrote that "a gold mine is like a tree with roots that spread in all directions." The laborers "find gold in outsize wells that they gouge out . . . very deeply[,] and when they find a gold vein, they follow it, digging to the end." He concluded: "No quantity of it is found . . . except through great effort and labor, and the cost of many people's lives, who perished in the mines."[34]

Portuguese merchants remained at a cautious remove from direct involvement in gold mining, continuing a hands-off policy that had begun in earnest after Lançarote's disastrous expedition in 1445. Upon reaching the coast near the gold fields, they left the coercion of enslaved people in mining operations entirely with the local Ashante and Dinkira leaders.

Despite episodic raids by other Europeans, Portuguese traders dominated Mina's gold trade for 155 years after the completion of the fort. In 1602, Dutch warships first attempted to oust the Portuguese, succeeding only on their sixth attempt in 1637. The new Dutch overlords kept the Portuguese custom of leaving the mines' operation and control of the enslaved laborers in the hands of these two Akan states. Willem Bosman wrote in 1705 that "the blacks know not to give us a notion [of the mines' locations], and none of us come so far inland, my friend, that's why I'm content with this." Bosman continued to discourage any efforts to alter the existing arrangements by invoking their present lack of access and the potential damage from violating a cultural taboo.[35] "Your Excellency knows better and that we have no access to the mines, and [I] don't even believe that any of us has ever even seen any . . . for the blacks regard this [gold area] as a shrine and will use every means whatsoever to keep others out."[36]

Whether such a shrine taboo existed, Bosman was keen to stress the hurdles a contest for direct control might bring.[37] Each of the native mine owners also jealously guarded the exact location of their wealth against potential rivals: "[The mining sites] are well guarded by their proprietors, that is by the rulers: for every ruler has his [own] mountains to mine, from which he has his people dig up the gold."[38]

The Akan mine owners also dominated the transportation of gold to the coast, leaving most of the coastal residents ignorant of the mines' locations. Two observers, a century apart, made nearly identical observations. Marees wrote in 1602 that "after investigating . . . I think there are few Negros who dwell near the sea, who know the mines' [location]," and William Bosman recalled in 1705 that "few blacks who dwell on the coast know [where] the mines are."

However, although the Portuguese largely avoided asking about the production of gold, their demand for it began to change the economic culture of the Gold Coast. The Akan leaders needed more labor than their community could supply to meet European demand. Here, the navigational skills of Portuguese pilots provided a solution that would allow them to avoid enslaving people from surrounding towns and villages, which would lead to conflict and the undesirable enslavement of people from expertise-enabled cities and towns. Instead, the Portuguese began transporting hundreds of enslaved Ewe, Aja, and Fon to a place so far from their homelands that escape was nearly impossible.[39]

The fort dubbed "São Jorge da Mina" did not survive in its original form. Following the Dutch takeover in 1638, the new landlords began to refer to it as "El Mina," often shortened to "Elmina," a form that makes no sense in either Iberian language but is far easier to pronounce in Dutch. It is still known as Elmina today.[40]

Numerous commentators have seen Lançarote's captures in the Bay of Arguin as the beginning of the deadly transatlantic slave trade.[41] After ordering their unjustified kidnapping, removal from Africa, and brutal separation from friends and family, a mounted, whip-wielding Prince Henry disposed of these people callously, to be exploited as laborers in a remote and unfamiliar land. With the advent of large-scale gold trading by the Portuguese, the intensity of that exploitation increased. Mines are among the most dangerous workplaces in the world and even today have some of the highest death rates of any occupation. These actions certainly foreshadowed the transatlantic trade decades later, during which captured people were sold to toil in deadly plantation agriculture. The malignant indifference to human life that defined the transatlantic slave trade began as Europeans and native leaders became used to treating human beings as expendable sources of profit.[42]

There were no heroes in this practice. Indigenous chiefs supplied the Portuguese with enslaved people, and the Portuguese, in turn, sold them to other native leaders who sent them to mine for gold, which they, in turn, could trade to Europeans for considerable profit. No one was more rapacious or deserving than anyone else. However, the Portuguese merchants who were satisfying the demand for gold and their European customers bear a significant portion of the responsibility for accelerating the growth in demand for slave labor.

During these years, Portuguese pilots undertook extraordinary risks and displayed tremendous skill in mastering the crisscrossing currents of the Gulf of Guinea. However, any celebration of their achievements in this realm must be tempered by a sober recollection of their objectives, which were as sordid as their navigational skills were outstanding.

A DEAD END?

Having sailed south of Benin, past the broad bayous of the Niger Delta, Portuguese ships soon halted again, unable to travel farther south. Like nearly all the world's seafaring vessels, Portuguese ships had only ever sailed north of the equator, under a sky in which their pole star, called Polaris or the North Star, seemingly stood still while a pair of bears rotated counterclockwise around it. Just south of the equator, a long, slender, finger-like peninsula—named for the Portuguese captain who first reached it—pointed its fingertip into the ocean, seeming to warn, "No farther." Just beyond the scolding finger of the Cape of Lopo Gonçalves, now called Cape Lopez, the pole star disappeared at two degrees south—Polaris would not guide them any farther. To continue, the Portuguese needed to accomplish two goals. First, they needed a new astronomical tool: the quadrant, which would revolutionize navigation at sea. Second, they had to map the stars of the Southern Hemisphere. These two technologies would open both hemispheres for travel for the first time using the same technology.

8

A Star to Steer Her By

I must go down to the seas again, to the lonely sea and the
sky,
And all I ask is a tall ship and a star to steer her by.

—John Masefield, "Sea Fever," 1902

As the sailor braves the dark waters at night, he may glimpse Masefield's star, a beacon in the heavens so steadfast it seems motionless. While other stars dance across the night sky, this star remains resolute. Seasons come and go, years pass, and yet the star remains unmoved—or so it seems. The North Star does shift, but its gradual movement is discernible only over a century. In a single lifetime, it appears not to move. The sailors' ingenuity in using this seemingly fixed point in the sky for navigation is remarkable.

Portuguese pilots likely used the North Star from their first voyages on this previously unexplored ocean. Dishearteningly, however, all records describing their methods for navigating with it have been permanently lost to that very same ocean. Only when their pole star appeared to be in imminent danger of disappearing below the horizon did the few surviving reports (from a pair of Italian pilots and a German printer) pause to describe their methods of using the North Star. These surprising pages tell us novel ways of using a star to steer by.

Three groundbreaking practices emerged among fifteenth-century sailors. First, they began to use scientific astronomical instruments to measure the North Star's height above the horizon. Second, they applied basic arithmetic to their instrument readings to measure distance from their home port. Finally, they developed the first instrument of universal navigation, usable not just where and when the North Star was visible but anywhere on Earth. This scientific revolution allowed Europeans to explore the world's oceans and paved the way for European colonialism, forever altering the course of history.

NAVIGATION BY FIXED STARS: THE PREAMBLE

The earliest records of using "a star to steer by" come from rival seafarers in the eastern Mediterranean. Around 1200 BCE, citizens of both Greek and Phoenician city-states began to transport goods by sea. To efficiently carry out their trade, inhabitants of the Greek city-states set up outposts in the Black Sea and began to sail north and east between their home islands and their Black Sea coastal towns. At roughly the same time, residents of the Phoenician cities set out in a different direction, traveling west across the Mediterranean to establish toeholds along the northern coast of Africa. Like their Greek counterparts, traders from cities such as Byblos, Sidon, and Tyre made regular east-west journeys to and from their North African outposts.

From the eastern end of the Mediterranean, Greeks and Phoenicians could travel away from land at dawn and dusk by using stars that always remained visible, like those near the North Pole. Hence, both Greeks and Phoenicians used the northern stars, but different constellations. The story of Greek wayfinding first appears in an epic poem describing how the legendary Odysseus observed the stars at sea,

> [keeping] his eye on the Pleiades and late-setting Boötes,
> and the Bear, to whom men give also the name of the Wagon,
> who turns about in a fixed place and looks at Orion,
> and she alone is never plunged in the wash of the Ocean.
> For so Kalypso . . . has told him
> to make his way over the sea, keeping the Bear on his left hand.[1]

Odysseus follows three constellations that sometimes disappear beneath the horizon (Pleiades, Boötes, and Orion) and a fourth, Ursa Major, that "alone is never plunged in the wash of the Ocean." In other words, Odysseus could rely on the Bear to remain visible. The nymph-goddess Kalypso had advised him that by keeping the omnipresent giant Bear to his left, he would travel in the right direction, namely to the east.[2]

In contrast with the Greek use of the giant Bear constellation, Phoenicians relied on a single star in the Little Bear. While they named it the "North Star," it was not the same star we use today. Their North Star, Kochab, lay in the same constellation but at the center of the Little Bear's body, while our North Star rests at the end of the Little Bear's tail. Why did they not use Polaris? When Phoenicians scanned the northern skies in the second millennium BCE, Polaris lay about sixteen degrees from the North Pole, making Kochab a more reliable guide. Curiously, on the other side of Earth, astronomers in China included Kochab, but not our modern Polaris, among the five central heavenly stars, called the "emperor," suggesting that at least one revision of their calendar occurred while Phoenicians glided through the Mediterranean. Since then, change has come very slowly.[3]

Every three thousand years, a different star comes closer to the North Pole and then moves away, only for another star to come closer, arriving after another three thousand years. Therefore, during the two thousand years following the rise of the Phoenician and Greek city-states, Kochab gradually drifted farther and farther away from the North Pole, leaving a space in the heavens where a pole star had once rested. At times—during the rise of the Roman Empire, for example—no star lay close to the North Pole. Eventually, the tail of the Little Bear came closer to the pole. By the start of the fifteenth century, when Portuguese navigators began sailing farther and farther south of Lisbon, Polaris lay more than three degrees away from the exact top of Earth. At some unknown time between 1455 and 1555, astronomers in Portugal made a series of calculations that allowed sailors to correct for the three-degree, twenty-eight-minute difference between the pole star and true north. Nearly 450 years later, Polaris has moved less than a degree away from true celestial north.[4]

REGULATIONS FOR THE NORTHERN STAR

As Portuguese ships first sailed south along the eastern edge of the Atlantic, they began using a north star differently than the Greeks or Phoenicians did. Rather than keeping it in sight to stay within the narrow width of the Mediterranean, they began using it to find out how far away they were from Lisbon. As the pole star threatened to sink below the horizon, leaving navigators without their time-honored celestial landmark, they increasingly faced the problem of how to navigate beneath a radically different sky.

By the mid-fifteenth century, the pole star appeared closer and closer to the horizon as sailors ventured farther south. Dreading its disappearance, two Italian merchants—one from Venice, the other from Genoa—described their fears after sailing more than fifteen hundred miles south of Lisbon. In a letter dated December 12, 1455, the Genoese merchant Antoniotto Usodimare wrote to his financial backers: "Although it will seem impossible to you, I believe that if you sail for one day more beyond Gambia you will lose the north [star]."[5] That same year, a Venetian merchant relayed identical concerns. While also anchored at the mouth of the Gambia River, the Venetian merchant Alvise Cadamosto advised that his crew had spotted the pole star just once there because it was so low on the horizon. "We only succeeded in seeing it," he wrote, "because the skies were apparent, and when we saw it, it appeared to be no higher than [the length of] a lance above the horizon."[6] Cadamosto's and Usodimare's concern arose when their ships reached the entrance to the Gambia River, just ten degrees south of the Tropic of Cancer and a considerable distance from the equator.[7]

Five years later, Portuguese captain Pedro de Sintra reached the Red Cape, where he noticed the pole star sinking closer to the horizon. The writer of this source did not care to report exact numbers, but he did note Sintra's remark that the star lay no higher than a man's height above the horizon.[8]

Another widely held but mistaken belief increased these merchants' anxiety. Usodimare mentioned a second reason he feared his ships were closing in on the equator: "The land . . . is almost at the equinoctial line . . . [because] in July the days are 12 and a half hours and the nights 11 and a half." Indeed, at least twelve hours of sunlight appeared at the Gambia River from mid-March until the end of September.

Although he had anchored more than eight hundred miles north of the equator, Usodimare's trepidation stemmed from the centuries-old practice of dividing regions from north to south on the basis of relative length of daytime hours. Since at least the time of Ptolemy, most geographers had believed that a traveler would have reached the equator when the hours of day and night were identical. A Portuguese pilot echoed this mistaken belief, declaring that "in this place, the equinoctial line begins because the days and nights are always equal."[9] Equally long days and nights result from a more complex interaction. This older and simpler determination lacked two critical factors: the day of the year and the sun's declination. While expressing their anxiety, these voyagers also revealed the novel methods they were employing, namely a scientific device for measuring the height of the pole star.[10]

The instrument they used to note the slow descent of the North Star was called a quadrant or quarter (of a circle). First used in India between 1500 and 500 BCE to measure a solar eclipse, it reached Ptolemy circa 150 CE.[11] During the Middle Ages, Muslim scientists constructed a variety of quadrants to carry out astronomical measurements and calculations. By the fifteenth century, scientists throughout Iberia knew that this precisely calibrated instrument could be a reliable device for nighttime observations of the heavens. Applying the device to sailing marked a significant innovation in Portuguese navigational techniques.

Two reports of quadrant use survive from 1460, both in documents that endured only because a fifteenth-century German printer took copies to his homeland. In one, Diogo Gomes bragged that he was the first to arrive on the island of Santiago in the Cape Verde archipelago. However, he then proceeded to undermine his self-congratulation with a second claim, boasting that "I had a quadrant when I went to these parts, and I wrote the heights of the pole Star on a map, and I found it [my measurements] better, than the [existing] map."[12]

Gomes could only have bragged about making superior measurements if he was comparing his actions to those of someone who had visited the islands earlier. Sure enough, Cadamosto had located the island of Santiago five years before. Thus, contrary to what he intended, Gomes gave away the probability that Cadamosto may have used a quadrant to find the North Star's zenith. If Cadamosto had made the calculations that Diogo Gomes claimed to have improved upon, then the quadrant would have already been in use by 1455.

Another document preserved by the same German printer recounts Genoese Antonio Noli's officially sanctioned voyage to determine the ease of sailing to the deserted Cape Verde islands. Following the southwesterly North African current at a mean speed of one knot per hour, Noli reached Santiago Island after only two days and one night at sea.[13] Upon reaching the main island of Santiago, he, too, employed a quadrant to measure the elevation of the pole star.[14] These events show that using a scientific instrument to register the North Star's position above the horizon began in the fifteenth century. Although they used the same instruments, pilots deployed them differently than landbound astronomers did.

Like all scientific quadrants, the ones carried on board were marked in degrees. Before leaving Lisbon, navigators scrupulously marked the maximum height of the pole star. Later, anchored in calm seas or on shore, they measured Polaris's height above the horizon with the quadrant. Then, subtracting the degrees of the star's local height from that at Lisbon, they multiplied the difference by sixteen and two-thirds leagues (an early estimate of the length of a degree of latitude). Thus, they calculated how far south they had sailed. Though imperfect, this process provided consistency in distance measurements from one voyage to another. So, if a previous pilot had reported a particular destination at 166 leagues south of Lisbon, and their calculations from the pole star said they had only traveled around 149 leagues south, they would know that they were sixteen and two-thirds leagues away from their intended harbor. Using a scientific instrument created a set of uniformly calculated data that allowed pilots to share information from prior journeys and introduced a new level of certainty for starting to sail in previously unexplored seas.[15]

While innovatively adapting a scientific instrument for navigation, this technique eventually ceased to be helpful. In the 1470s, ships had to sail south of the equator to return from the Guinea coast. Soon, pilots would have to abandon their reliance on Polaris. By the end of the 1470s, Portuguese ships had advanced only as far south as Cape Saint Catherine. The customary explanation for their failure to proceed further relies on a geographic limit that royal authorities had placed. However, a more compelling explanation came from the natural world; having reached two degrees south of the equator, they found that Polaris had drifted entirely below the horizon at the latitude of Cape Saint Catherine, leaving pilots uncertain how to continue their journey south.

A NEW STAR AND NEW SKIES

If they were to make any progress toward the desired passageway to the Indian Ocean, navigators would need to find another method for tracking their position. The most obvious option would be to find a southern pole star. However, there was one problem with that idea.

By the time ships had reached thirteen degrees north, they could already see the stars of the Southern Hemisphere. Near the Gambia River in 1455, Cadamosto produced the first surviving description of newly visible southern stars. "We also saw six clear, large, brilliant stars low [on the horizon]. Noting their location by the compass, we saw that they lay directly to the south, arranged in the following manner . . . which we judged to be the southern chariot."[16] Four of the stars in the chariot would later form part of the constellation called the Southern Cross. But Cadamosto continued to scan the newly visible horizon, searching vainly for a southern pole star. Disappointed, he wrote that "we did not see the principal Star." Nonetheless, he clung to the belief that such a star would become visible if only he sailed farther south. He did not press his luck, however: "It did not seem reasonable for us to discover it [farther south] because to do so, we would lose sight of the North Star."[17]

But Cadamosto's "principal Star" did not exist. No matter how far south any ship went, none would ever find a southern pole star.[18]

At this point, Portuguese pilots faced a conundrum. Upon accepting that there would be no South Star, they sought an alternative during the 1470s and the first years of the 1480s. In a remarkable and unprecedented move, the astronomers advising the pilots eventually recommended using a wholly different star: the sun, since it would be visible both north and south of the equator. Their recommendation led to the invention of a uniform navigation method that would work in both hemispheres.

The decision to switch to the sun rather than the moon or other celestial objects stemmed from a unique orbital characteristic of the sun. Every day, the sun advances in longitude by about one degree, corresponding to one degree of right ascension, meaning that the sun's predicted position will differ for every day of the year. The Portuguese found that a precise latitude could be calculated with the date, observation of the midday height of the sun, and the sun's declination.[19]

REGULATIONS FOR THE SUN

Anglophiles claim that the English achievement of devising a reliable means to calculate longitude was far more challenging than learning to calculate latitude, but the opposite is true. Unlike the solution for longitude, which meant devising seagoing clocks, finding latitude required an original mechanical instrument, the ability to predict the sun's passage through the ecliptic, and solving over a thousand complex trigonometric equations.[20]

Before sailors could be safely sent into the Southern Hemisphere, astronomers had to decide precisely what pilots needed to know to determine their position. They concluded that the navigator required access to three pieces of information: first, the correct date (month, day, and year); second, the maximum elevation of the sun above the horizon at midday, as determined by the navigator; and third (most difficult of all), the sun's exact position in its apparent annual path around Earth on that date. By combining these three factors, any pilot could determine the ship's precise position in degrees and minutes of latitude rather than merely the distance from home.

Using the sun as the principal celestial object vastly increased the complexity of the mathematics required for navigation. Instead of the simple arithmetic performed by pilots, land-based mathematicians and astronomers had to carry out complex trigonometric functions in advance. These calculations predicted the annual path of the sun, following a shape called an analemma, which resembles an exaggerated infinity sign or distorted, sideways figure eight.

However, the sun's journey along this figure eight shifted slightly each year. Thus, astronomers had to calculate more than a single annual table; they had to compute the sun's daily position for every year that followed. The first group of such calculations, likely initially worked out by José Vizinho, established the pattern for building future tables by taking four years, including a single leap year, and predicting the sun's path along the ecliptic for every day of those four years. That meant that 1,461 separate calculations had to be made and verified.[21]

A secondary reason for the large number of equations lies in their intended use. The predictions had to cover more than one year, because sailing from Lisbon to any location south of the equator and returning would take more than a year. With access to calculations made for only one year, pilots' records of their position would be significantly wrong by the time of their return voyage.

During the 1480s, the first three successful voyages south of the equator tested the new system. Armed with basic tables, calendars, and an unnamed instrument, Diogo Cão (pronounced "cow") led the first two successful journeys into the Southern Hemisphere, the second of which took him as far as twenty-two degrees south. Four years later and some seventeen degrees farther south, Bartolomeu Dias finally reached the long-fantasized and hoped-for sea passage to the Indian Ocean. Unfortunately, no information survives regarding the instruments Cão or Dias used to measure the sun's height above the horizon.

Almost a decade after Dias returned, Vasco da Gama set forth on the first successful sea voyage to the west coast of India and back. By the time of da Gama's epoch-making journey in 1497, scientists in Lisbon had successfully modified the astrolabe, another astronomical instrument with a lengthy and storied past, for use in imperfect weather conditions. By reducing its size, replacing the formerly solid backing with openings designed to allow moderate winds to flow through, and adding weight to the bottom for stability, they adapted the device for maritime journeys. Finally, to lower the potential for observer error, they cut back the number of degrees on the device to ninety on opposite sides (the top right and bottom left quadrants), so that a pilot measuring the sun's altitude could double check his readings on both sides. As with traditional astrolabes, a small but accurate movable arm containing two pinholes allowed pilots to rotate it to align the openings with the sun's rays. Thus, Portuguese scientists created a precise device for measuring the sun's zenith anywhere a ship could anchor. They also added built-in safeguards against observational errors by making repeated measurements before and after noon.

The introduction of this new method altered the unit of measurement. While using the quadrant in previous decades, sailors measured degrees by distance from Lisbon. With the novel device and different techniques, navigators began noting their position in degrees of latitude, thus dividing Earth's surface into coordinates. In a revolution only half understood at the time, they had devised a mathematical system that could accurately locate any place on Earth.

Finally, and not least, the Portuguese transformed navigation by inventing a means of fixing latitudes accurately anywhere around the world. The instruments used to measure the sun's midday height improved over time, but the fundamentals remained the same. Fixing precise degrees of latitude while employing a long-standing, less precise estimate of longitude called the "lunar distance model" would prove a remarkably successful method for over three centuries. It enabled the Portuguese and future European sailors to circumnavigate the globe and sail to all parts of the world for those three hundred years, thus unintentionally creating the means to establish European overseas colonialism.[22]

The first four voyages south of the equator marked a distinct shift in official policies. None of these initial exploratory expeditions involved abductions from unfamiliar territory. Dias alone returned to Elmina to acquire enslaved people, but he was promptly sidelined for nine years after his return. Instead, all four voyages had memorable encounters with at least one community previously unknown to Europeans. The ensuing voyages by Cão, Dias, and da Gama all involved calculation of latitudes, encounters with a range of previously uncontacted peoples, and views of some of the most spectacular landscapes Europeans had yet seen.

9

The Deepest River and the Oldest Desert

1480–1486

Crossing the milestone of the equator, a feat that revived the Portuguese dream of sailing around Africa, was a pivotal moment in maritime history. Flush with gold from the Guinea coast and armed with a new strategy for sailing without the North Star, Portuguese explorers could focus primarily on gathering nautical knowledge for this ambitious venture.

Three such expeditions stood out for the remarkable quantity of information they acquired about the South Atlantic. All took place during the 1480s, under the command of captains with extensive experience sailing along the Guinea coast. The first two journeys were led by the roguish Diogo Cão and the third by the dictatorial Bartolomeu Dias. While the latter's journey remains better known, those of Cão brought him to incredible natural wonders, like the world's deepest river on his first voyage and its oldest desert on his second.[1]

THE WORLD'S DEEPEST RIVER

Cão stopped at Mina for provisions before going on to Cape Lopez, which, according to him, lay one degree south of the equator. The shores north of the cape were dotted with rivers and harbors. However, lagoons and mangrove swamps prevailed just to the south, offering neither anchorage nor fresh water, so he continued south.[2]

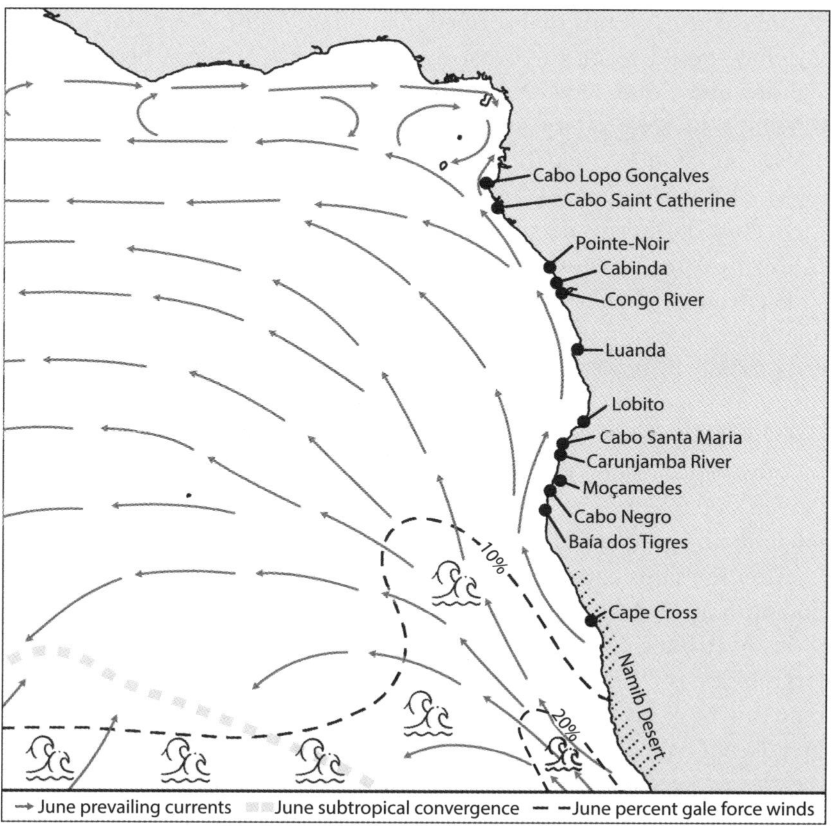

MAP 10. Coastal areas on Diogo Cão's journey. © Esri, Tomtom, Garmin, Open Street Map Contributors, GIS User Community. Data from UN Environmental Program, *World Atlas of Desertification*, 1997; U.S. Hydrographic Office, *Pilot Charts North Atlantic*, 2002.

As Cão's ships traveled southeast from the equator along the coast of today's Gabon, they had the good fortune to find an offshore current moving steadily southward. While another, narrower current moved northward, an occasional turn of the helm would quickly resume their southward journey. In addition, gentle breezes blew from the south and southwest, slowing their progress as they tacked back and forth. Still, the journey would have been smooth and straightforward, allowing careful observation of harbors and settlements along the coast. Only occasionally would nighttime storms threaten the voyage.[3]

As Cão headed south in late fall, intense tropical rainstorms materialized sporadically. Fortunately, early warnings appeared in towering cumulonimbus clouds that formed inland late in the afternoon, indicating they would reach the coast by evening. The interval between first sighting the clouds and the arriving downpour provided enough time for ships to seek safety in case the storm proved dangerous. These advancing clouds could bring an inconvenient hour-long squall or a night of torrential rains accompanied by fifty-knot gusts. However, these episodic storms disappeared by dawn, allowing Cão's ships to sail calmly near the coast during the day.[4]

Between patches covered by low clouds, the sailors would have seen mile after mile of impenetrable forests along the shore, interrupted only occasionally by beaches, lagoons, and small rivers that provided an opportunity to replenish their supply of fresh water. Significantly, few people resided along the coast beside these dense woods. Most residents survived by subsistence fishing and possessed few goods to trade. Duarte Pacheco Pereira said twenty-five years later, "Most of this land was unoccupied or inhabited by people with scant interest in trade or little to offer."[5]

After the ships had sailed past two hundred miles of shoreline without finding bays suitable as harbors, the coastline began to be marked by a series of striking red cliffs. Suddenly, they glimpsed what seemed to be a large black rock in the distance. As they approached, they discovered that this rock lay atop a densely wooded peninsula that protected the first decent harbor they had found since passing the equator. Upon entering the protected bay, they found the beach disappointingly deserted. Nonetheless, the water abounded with tuna, bass, sole, and sardines, and a small river north of the bay supplied fresh water. Hoping to find inhabited stops later, they sailed on. After seeing only heavy surf to the immediate south, Cão returned and marked the location of this bay by planting a five-foot (1.63 meter) limestone pillar atop the highest point of the black rock, where it would be easily visible from the sea. Showing a stupefying lack of imagination, Cão or his crew named the peninsula Pointe-Noir (Black Point). Because it remained unoccupied for the next four hundred years, the place retains this designation on maps, even after the arrival of commercial shipping in the twentieth century.[6]

Departing from Black Point a second time, Cão's small fleet continued southeastward, following the same easy currents and light winds. The rough surf he had encountered initially continued for the next twenty-three miles of the coast until they reached a small river leading into a lagoon, where the coast changed to hills. Thirty miles later, they ran into a second beautiful, but more petite, harbor with high cliffs surrounded by green hills and valleys. This lovely harbor, called Cabinda today, remains suitable for small cargo ships, but the colliding currents and swells made for occasional heavy rolling of Cão's vastly smaller wooden vessels.[7]

In mid-April 1484, less than ten miles away from Cabinda, the crew could have glimpsed about one foot (0.3 meters) of fresh water floating above the typically salty Atlantic. Any experienced sailor would have surmised that a river lay nearby, but Cão's fleet would continue southeast, going thirty-six miles farther before encountering the source of this water.[8] What they found was no ordinary river. Upon first viewing it, they were astonished as the previously crystal-clear ocean turned a muddy olive brown. Soon, they felt the river's force pushing them out into the sea, as the Congo River poured out 2.2 million gallons every second, three times what crashes over Niagara's largest falls in an entire day. The sheer power and magnitude of the Congo River would have left them in awe.[9]

Having sailed more than four hundred miles south of Cape Lopez, Cão had come upon a natural wonder. The crew's surprise at the size and power of the Congo River would have been the same had they arrived from anywhere in the known world: Europe, Asia, or elsewhere in Africa. Nowhere else had anyone ever seen a river blast such immense quantities of fresh water into the sea.[10] And yet, a mere thirty years later, Europeans would encounter another river that surpasses it: the Amazon, in the as-yet-unexplored New World. Even today, the Congo River remains second only to the Amazon in the size of its drainage basin. However, the Congo differs from the Amazon in one respect: its navigability, opening up the potential for further exploration and discovery.

The Amazon traverses Brazil's interior across broad, flat land broken only by scattered, sometimes large clumps of vegetation. However, 220 miles (354 kilometers) inland, the Congo widens into a lake called the Malebo Pool, where today the capitals of the two Congos sit on opposite shores, within sight of each other: Brazzaville, capital of the Republic of the Congo, on the northern bank, and Kinshasa, capital of the Democratic Republic of Congo, on the southern side. On the western side of the Malebo Pool, the river plunges through thirty-two separate cataracts over the next two hundred miles, rendering it completely impassable. This final section of the Congo River is the deepest riverbed on the planet.

On a return voyage to Congo, Cão sought to travel upriver, but any hopes that he might connect with other communities, or even perhaps the Indian Ocean, were dashed. Ninety-three miles (150 kilometers) inland, he encountered the rocks, whirlpools, and islands, making further navigation impossible. To mark the end of his efforts to sail up the interior of the Congo River, he carved giant graffiti on a massive black stone beside Lelala Falls: "The ships of the enlightened King Dom João the Second of Portugal arrived here." To the right, he inscribed his name and those of his fellow leaders: Pedro Anes and Pedro da Costa. On another rock farther to the right, he carved the names of six crew members: Álvaro Pires, Pero Escolar, Antão (possibly an enslaved person), João de Santiago, Diogo Álvares, and Gonçalo Álvares. The six-hundred-year-old graffiti can still be read today.[11]

To create a visible landmark for future ships, Cão ordered a giant square stone to be positioned atop a pillar on the south bank of the river, on a point that extended farther into the Atlantic than the northern bank.[12] This placement allowed boats to see the southern side of the river from miles away in either direction. Because of the pillar's visibility, the south side of the mouth of the Congo River was labeled Pillar Point (Punta do Padrão) on many of the region's early maps.

While this and later pillars served as navigational markers, their origin has remained unexplained. One clue to their creation lies in a design proposed in 1452 when the Portuguese ambassador suggested erecting a pillar to mark Afonso's sister's marriage to Frederick III, Holy Roman Emperor, in front of the church where they wed. The pillar was to be "a very tall marble column with the Royal shield of Portugal."[13] A nearly flat square stone was perched atop the pillar, inscribed on both sides with the royal shield as well as the couple's names and the date of their marriage. According to King Afonso's chronicler, the location of the pillar was "in memory of this first view, in the very place where they [the couple] were first seen,"[14] thus commemorating the location of an initial view.

For the next fifteen years, all initial voyages south of the equator carried similar stone pillars and a square stone with the royal shield to be placed on top. Designed in Portugal, these navigational landmarks may have been meant to signify the first Portuguese view of a coast, just as the pillar constructed outside the church commemorated the initial appearance of a couple. The only addition to this design was a small stone cross that had to be attached while on location.

However, on his initial journey, Cão explored only the entrance to the Congo River. There, to his delight, he encountered a remarkable civilization with well-watered, fertile land; large-scale agriculture; and social hierarchies as complex and impressive as any already found in the Gulf of Guinea.

As on nearly every previous journey, Cão's fleet had several enslaved translators from lands they had previously encountered. However, these interpreters did not understand a word of Congolese. Hence, the two sides resorted to gestures. While such physical communication often results in inadvertently offensive gestures and misunderstandings, this encounter proved surprisingly gratifying.[15]

Cão sent several of his crew into the interior, accompanied by local residents who led them to meet their ruler, who resided some distance inland. Cão waited over three months for their return, exchanging cloth for supplies from the inhabitants. Perhaps growing impatient, he observed that many locals visited the ship readily and so decided to take four of the ship's regular visitors back to Lisbon, so that the Portuguese could learn Congolese and vice versa, opening the door to mutually beneficial exchanges. By September and the beginning of spring, his emissaries had yet to return, so Cão communicated via sign language that he was leaving his as-yet-unreturned crew as proof that he would return with the four men. Cão promised a return in "fifteen moons," the common fifteenth-century expression equivalent to "a couple of weeks." How or whether the locals understood that expression while still using sign language remains unclear.[16]

At the end of September 1483, Cão sailed back to Lisbon, where King João II was pleased to meet such well-mannered and intelligent men and the news of a vast and prosperous realm. As he had promised a swift return, Cão returned to the Congo River by the winter of 1484 to rescue his crew, loaded with gifts for the ruler of the Congo. Upon his return, the locals were pleased to see their fellows return alive and treated well.[17]

On this return trip, Cão finally met with the ruler of the Congo. At this point, his shrewd choice to show off Lisbon to a quartet of Congolese men began to pay off. The lord of the Congo agreed to send a contingent of his people for a more extended stay in Lisbon, including his courtier Caçuta and the sons of several leading members of the community, who were to be baptized. The younger men were also to be educated in reading, writing, and religion in monastic communities, where they would impress their teachers with their intellect and manners over their two-year stay. This custom of educating the sons of overseas elites eventually became a staple exercise of "soft power," used for centuries by many other empires to cement cultural authority overseas.[18]

The Congolese leader realized the significance of baptism as the central initiation rite of Portuguese society. Hence, a few years later, when seeking military assistance from the well-armed Portuguese, he insisted upon immediately being baptized before embarking on an urgent military campaign. As a result of this deft bit of diplomacy, Portuguese soldiers accompanied the Congolese chief in putting down a rebellion by his enemies, the Bantu-speaking Teke people, who had moved southward and occupied islands in the Malebo Pool.[19] This campaign proved the cultural and strategic political advantages of providing armed assistance to native allies. Eventually, it became a central strategy as the Portuguese advanced inland in other territories, including the undiscovered land of Brazil.

While Cão, his men, and later the Portuguese king and his court treated Congolese leaders with respect and wrote admiringly of their manners and intellect, their praise for this community concealed another agenda. The Portuguese perceived that an African empire of this scale and complexity would be able to deliver massive numbers of enslaved people from the communities they had subjugated. However, owing to an absence of significant conflicts that would readily generate exchangeable prisoners of war, sales of ivory flourished as the slave trade initially developed more slowly than the Portuguese had anticipated.[20]

For now, the Portuguese would have to look elsewhere for human cargo. While Cão's first return mission was partly centered on returning the four Congolese to their homeland with tales of the wonders of Portugal, his second assignment required exploration of the coast to the south of the Congo River. Before departing, Cão left three men behind to learn the language and inform themselves about a possible trade, promising to collect the trio on his return.

FROM THE CONGO RIVER TO THE SKELETON COAST

Having left this small contingent behind on the Congo River, Cão began following the coast southward, still searching for that elusive connection to the Indian Ocean. On this second journey, he found no other river as mighty nor any civilization as sophisticated as the Congo's. However, he succeeded in identifying the two best harbors in the South Atlantic, including its most abundant fishing grounds. He ended the journey there, becoming the first European to see the coast of the world's oldest desert.

Initially sailing south from the mouth of the Congo River, Cão passed a broad delta twenty-five miles long where the river weaved in and out among numerous large islands.[21] For the next 150 miles, the crew sighted many small rivers nested between cliffs and hills, suitable for resupplying the ship with fresh water but lacking significant natural wonders and with few, if any, coastal dwellers. The sailing was perhaps not as smooth as it had been from Cape Lopez to the Congo, because the offshore current occasionally reversed direction, but generally the expedition continued southward at a leisurely 0.8 knots.[22]

Having covered less than half the distance between the equator and Congo, Cão came across clumps of trees interspersed between picturesque pink granite hills. Just to the south, he spotted a beautiful bay, known today as Luanda and still considered the best harbor in the eastern South Atlantic. The natives raised goats on two islands off this harbor, leading to the Portuguese naming the place Island of Goats (Ilhas das Cabras). More importantly, along the shore resided a beautiful half-inch marine snail called "nzimbi" (*Olivella nana*), which was used as currency all along the coast and throughout the kingdom of Congo.[23]

This snail's shell became a key currency for the Europeans as well. A 1506 visitor reported that fifty shells would buy a hen, while a goat would cost three hundred. Although lacking in significant trade goods, the prominent local Bantu-speaking population raised plenty of chickens and goats for which the crew could trade, establishing the bay as an essential future resupply depot.[24] Thus, well equipped with food and fresh water for their onward journey, the small fleet departed southward, likely having either captured a local or persuaded one or more Ambundu residents to accompany them, again to serve as future translators.[25]

The next long section of the coast resembles a giant serrated bread knife, poised at an angle, with small, tooth-like capes ready to slice into the Atlantic. These sharp edges cut deeply into the sea, leaving neither a continental shelf to follow nor shelter against the winds and waves from the north. After passing two hundred miles of this knife-like coast, Cão spied a long, narrow bay protected from the sea by a sand spit. Since the fishermen caught fish in tide pools, Cão called it "Tide Pool Point." This haven, now called Lobito, remains the second-best sheltered harbor in the eastern South Atlantic, after Luanda. By then, Cão had journeyed four hundred miles beyond the mighty Congo.[26]

However, once beyond Lobito, shelters became fewer, and the coast changed direction, dismaying the crew. The steady, gradual southeastward progress since leaving the Congo River halted as the coast veered sharply into the Atlantic. Anchorages became few and far between. Because the shore held few sheltered bays, Cão had to avoid being beached by the swells pushing vessels toward the coast.

The coastal environment continued to shift as the sailors continued southwestward. At first, the alterations were subtle. Small rivers still emptied into the ocean, with tree-covered hills on either side. However, between these tree-lined rivers lay gravelly hillsides partially covered by patches of shrubs and stubby plants. In 1507, Pereira laconically declared, "There are not as many trees as there were previously."[27]

One hundred miles southwest of Lobito, Cão's small fleet reached a small, sheltered bay called Cape Saint Mary. Although only a mile wide, this uninhabited bay furnished fresh water and plentiful fish for the onward journey. After briefly departing to the south, Cão spotted a distinctive rocky hill just east of the cape. It resembled a woman's hat with a flat top and rounded base. Since he was charged with setting up pillars at readily identifiable points along the coast, he turned his ship around and erected another pillar as a reference point for future voyages. At this point, Cão was just over five hundred nautical miles south of the Congo River.[28]

For the first thirty miles south of Mary's Bonnet, the shore consisted of a chain of dark gray cliffs resembling a bracelet whose links had all been stretched open and separated by sandy beaches. Inside these broken links lay small rivers and bays. The bracelet vanished as Cão arrived at Carunjamba, a large seasonal river with plenty of potable water beneath the riverbed during the dry season (June–September). Once past this river, bright pink-orange sand appeared: the Namib desert, often considered to begin here, continues to the Olifants River in modern South Africa. Unbeknownst to Cão and his crew, they had reached the northern edge of a long, unseen coastal desert. This increasingly dry coast accompanied a change in ocean currents.[29]

Since leaving the Congo River, Cão had been sailing along the Angola Current, which changed direction every few weeks, two months at most. While it often flowed southward, in the Southern Hemisphere's autumn (March–May) its direction became unpredictable. Indeed, the only dependable characteristic of this current was its general but unpredictable southward trend. As a result, the ships could often sail faster—shortening their journeys by weeks to months—while heading in that direction, but the boat would suddenly slow to a crawl when the current began flowing another direction.[30]

When the current proved contrary, Cão would have needed to rely on the daily reversal of temperatures over sea and land. Since land heats and cools faster than the ocean, breezes blow from sea toward land during the day. After dark, the land cools faster than the water, and light winds blow toward the sea. When close to shore, Cão's ships would have sailed during the coolest hours of the day. However, these nighttime breezes could begin between sunset and shortly after sunrise, forcing crews to remain watchful for their sudden onset.[31]

In the stretch of coastline 30–120 miles south of the pillar at Cape Saint Mary, the erratic but usually southward-heading Angola Current encounters the cold Benguela Current rushing up from Antarctica. Oceanographic research has shown that the Angola and Benguela Currents have squared off against each other in this space (between fourteen and sixteen degrees south) for thousands and possibly millions of years.[32] Unlike the inconsistent Angola Current, the northern Benguela Current travels reliably northward at 0.5 and 0.8 knots year-round. But it also produces unwelcome surprises. Northbound waves over twenty feet (six meters) appear suddenly and without warning, making it one of the few ocean areas where steep waves travel in the same direction as the wind.[33]

Along with the new current's altered direction, the winds also shifted. The Angola Current had advanced on the outer edge of a clockwise-moving low-pressure system, assisting Cão's journey southward. However, just south of this low-pressure front, a high-pressure zone usually appears with winds that blow in the opposite direction, namely counterclockwise. When he reached this subtropical high pressure, Cão would have sailed against winds blowing toward him from the south and southeast at eleven and sixteen knots.[34] While these opposing winds grew stronger as he continued south, his second major environmental challenge was about to appear.[35]

A few miles south of where the Angola and Benguela Currents collide, Cão's fleet came across the Carunjamba River. The river follows a fast course through rocky terrain but reaches the ocean only during the rainy season. South of the Carunjamba River, the landscape grew even drier; the areas of shrub-topped hills shrank, separated by vast expanses of sandstone and desert. Over the next seventy-five miles, the crew found fresh water at only two small, seasonal rivers. During the dry season, these rivers did not quite reach the sea, requiring them to dig into the sand for fresh water.[36] Although he did not realize it then, Cão had reached the northern edge of a desert that continued for hundreds of miles, ending only two hundred miles (320 kilometers) north of the Cape of Good Hope.

Cão might have hoped the desert had ended when he encountered a large bay backed by gleaming white cliffs. While not as sheltered as Luanda or Lobito, these waters abounded with sea life, attracting two substantial native settlements. The presence of two such villages led Cão to name the bay "Anchorage of the Villages." Since then, this area (now known as Namibe or Moçamedes Bay) has become the principal South Atlantic fishing grounds.[37] However, any sliver of optimism Cão might have had about finding further settlements would soon be smashed, as south from Moçamedes the land would become even drier. For the next 460 miles, Cão found either harbors without water or rivers with water but without shelter.[38]

Thirty miles after leaving the Anchorage of the Villages, Cão spotted a striking headland. Since he situated pillars atop landmarks readily visible from a considerable distance at sea, Cão planted a stone column atop this headland two hundred feet (sixty-two meters) high, composed of weathered black rocks surrounded by sand and low shrubs, which Cão unsurprisingly labeled Black Cape (Cabo Negro). As the pillar's isolation kept it safe from humans, its text remained legible over a hundred years later, when it was found and returned to Lisbon, despite centuries of exposure to the elements.[39]

Over the next hundred miles southwest of the pillar atop Cabo Negro, the scrub-covered areas shrank even further until they vanished entirely, replaced by a barren landscape of rock ledges whose tops had been flattened by winds and waves separated by sand dunes and expanses of gravel.[40] Eventually, even these rocks and gravel vanished too, replaced by mile after mile of dismal yellow-gray sand. When Cão passed the rich fishing grounds at Tombua, he might have seen fishermen who had set up temporary huts of curved whalebone ribs covered with seaweed mats, where they could hide from the sun, despite the absence of fresh water nearby.[41] In the mostly forbidding terrain, a dismal landscape where offshore fish again abounded, Cão would discover only two other decent harbors. Desert sand surrounded both of those bays, however, and fresh water lay miles away. Just south of Tombua lay the uninhabited Rock Point, named for the scattered black rocks found among the sand there. Today it is known as the long sandy peninsula that creates the Bay of Tigers (Baía dos Tigres).[42]

A little less than forty miles south of the desert that surrounds the Bay of Tigers, Cão came across another sizable seasonal river, called the Kunene.[43] To his relief, the coast's direction shifted, once again heading southeastward toward the Indian Ocean. However, finding fresh water and a change in the direction of the coast remained the only welcome news. South of the Rio Kunene lay nothing but desert and an experience unique in human history.

THE WORLD'S OLDEST DESERT

This next strip of the Namib desert has become infamous since Cão's initial visit. Because scores of whitened skeletons jut from beneath the dunes, it has become known as the Skeleton Coast, originally for the remains of massive whales who perished along its shores.[44] No humans had ever inhabited this bleak coastal desert. Between two and three millennia ago and several hundred miles inland, a small area experienced rainfall, allowing a small settlement to exist until 1000 CE, when the rains stopped. This tiny community had no reason to enter a deadly desert that had remained dry for millennia.[45]

However, along the shore, few animals and fewer plants survived—only one peculiar succulent and a few lizards and beetles. This place had become a desert when modern mammals first began to roam the earth, and Antarctica cooled from a tropical to a temperate forest. By contrast, the more familiar Sahara has existed for only two to three million years. But the Skeleton Coast had survived for over fifty-five million years. Yet despite its age, Cão and his fleet were the first humans to have seen this coast, the first glimpse of an unwelcoming ancient land.[46]

Sailing along this coast would have been excruciating at some times and deadly at others. Winds could blow the coastal sand far out to sea, diminishing visibility and irritating the skin like airborne sandpaper.[47] For half of the year, though, the hot desert air prevented the cold ocean air from rising and producing rain, resulting in a thick, damp layer of fog that blanketed the coast. Blown inland as far as fifty kilometers, this fog concealed the deadly sands from view, resulting in the hulks of ships and human skeletons later joining the whale bones buried in the sands.[48]

As he continued southward along this unique and ancient desert, Cão battled an even faster Benguela Current. With the coast obscured by fog, he would have needed to stay several miles from the coast, perhaps tacking back and forth in the face of oncoming winds or relying on the alternating land and sea breezes.[49] Either way, his southward journey would have been arduous and his progress slow. But more significantly concerning was the lack of fresh water. Only two places might have possibly provided fresh water past the Kunene River. Rain occasionally falls far inland, where water drips into the sands below. Sometimes, that deeply buried water may be detected visually as a slightly indented area of sand, but nothing guarantees that the water is still there. Cão's fleet might have been able to find water by digging deeply beneath a flat sandy area, but there would be no guarantees it had not already evaporated.[50] For nearly three hundred nautical miles beyond Kunene, Cão would find no visible fresh water.[51]

PLACING THE FINAL PILLAR

After venturing nearly a thousand miles beyond the Congo, Cão had reached a turning point, confronted by two distinct environmental obstacles. Strenuous sailing demanded large quantities of fresh water for the crew, and since almost none could be found along the Skeleton Coast, supplies on board had fallen dangerously low. Secondly, more potent winds from the Southern Ocean made further southward progress difficult. By twenty-one degrees south, gale-force winds would have blown from the Antarctic 10 percent of the time, pushing Cão back north.[52] Just south of the final pillar left by Cão, ships would begin to encounter stronger winds even during late spring and early summer (November–February), otherwise the best time to sail.

Seeking a good site for a pillar, Cão came across a rocky point three miles from the coast. He planted the stone pillar, weighing 1.1 tons and standing eleven feet (3.5 meters) tall, in the sand among the scattered, weathered black rocks.[53] The spot eventually became known as Cape Cross.[54] Surrounded for centuries by a large seal community, the pillar remained intact and partly legible until it was seized by a German explorer, who brought it home in 1893. There it remained until 2019, when it was returned to present-day Namibia.[55]

Sometime during his return to Lisbon, Diogo Cão died of unknown causes, at the age of thirty-six.[56] Fernando Pessoa wrote of the final pillar Cão planted:[57]

> I, Diogo Cão, navigator, left
> this pillar at the foot of the brown sand
> And onwards I sailed. . . .
> This pillar advises the wind and the skies
> That of this daring effort, my part is finished.

Until Vasco da Gama's historic voyage, Diogo Cão's journey was considered, throughout Europe, the most daunting maritime achievement of the era, eclipsing that of his now better-known successor, Bartolomeu Dias. According to calculations made in Lisbon, Cão had nearly reached the sea lane to the Indian Ocean.[58] Carmen Radulet writes that in Europe during the fifteenth century, "the fame of Diogo Cão far surpasses that of Bartolomeu Dias."[59] Even during his lifetime,

Cão was held in far greater esteem than Dias, for he received the honor of being made a member of the nobility. Only da Gama received greater recognition.[60]

The significance of Cão's ennoblement is often underestimated. It conferred immense social superiority over other members of society, no matter how successful the latter might be. In his decree ennobling Cão, King John II asserted that he was "actually differentiating . . . and separating [Cão] from inclusion in the plebians [by] authorizing *his creation as a noble* with a coat of arms."[61]

By custom, titles were rewarded only for military valor, and not for the navigational feats and cultural contacts that Cão had made. To ensure that Cão would be accepted within noble circles, the king's staff had to scour his family history to locate a military pretext for awarding the title. Therefore, after mentioning his achievements, the king reached far back to when one of Cão's ancestors had fought alongside the king's great-great-great grandfather.[62]

With this necessary affirmation of historical military valor inserted into the decree, King John II ensured that Cão and all his legitimate descendants would be recognized as rightfully titled. As a customary part of this ennoblement, the king ordered the all-important registrar of heraldry to draw up Cão's coat of arms. Dias received no such recognition during his lifetime or for centuries after.

10

A First Glimpse of the Indian Ocean

1486–1488

Following the untimely demise of Diogo Cão, Bartolomeu Dias immediately embarked on a historic voyage south of the equator. He commanded two additional ships and a supply vessel for this journey, leading the first-ever expedition from the Atlantic to the Indian Ocean.[1]

Despite this accomplishment, Dias received no significant acclamation during his lifetime because of his headstrong decision-making, flawed and often careless judgments about suitable harbors, and placement of navigational beacons.[2] Despite evident navigational skills, Dias's problematic leadership also produced unnecessary conflicts with his crew and the natives he encountered. Following his return, the king repeatedly snubbed Dias, deliberately keeping him away from any crucial voyages for the next nine years. Only when Pedro Álvarez Cabral's massive (twenty-seven-boat) fleet sailed to India in 1500 did the Crown finally allow Dias to participate, but only captaining a single ship and far from any leadership role. Near the end of the fleet's Atlantic crossing from Brazil, Dias and his boat sank off the Cape of Good Hope, almost exactly where he had previously first succeeded. Whether this failure signified that his initial rounding of the cape had been a significant achievement, a fluke, or just good luck is hard to say.

Dias set out from Portugal with three ships on the second attempt to sail beyond Diogo Cão's final pillar. The previous attempt failed because the desert continued for another thousand nautical miles beyond Cape

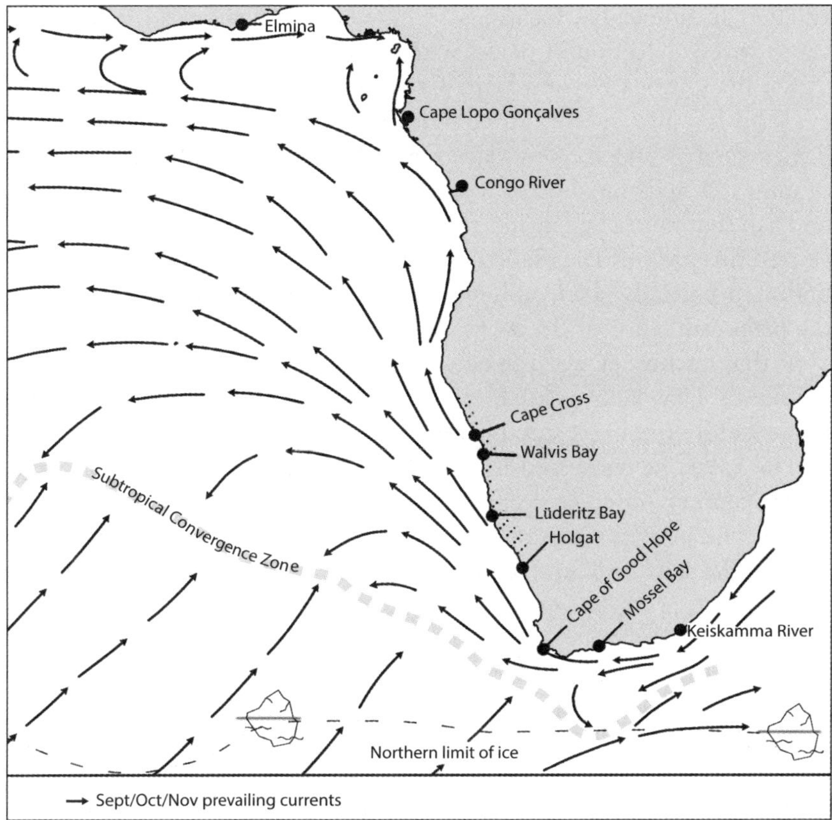

MAP 11. Coastal areas on Bartolomeu Dias's journey. © Esri, Tomtom, Garmin, Open Street Map Contributors and GIS User Community. Data from U.S. Hydrographic Office, *Pilot Charts North Atlantic*, 2002.

Cross. Consequently, the crew ran out of fresh water and was forced to turn back. As a result, Dias's three vessels were accompanied by a supply vessel containing water and food to ensure the crew's survival. This practice became routine on long voyages after proving its value on Dias's voyage.

In 1486, likely during October and November (spring in the Southern Hemisphere), Dias traveled south across the equator, heading to the final pillar Cão had placed. During those months, the rough seas had retreated 180 miles south of Cape Cross, allowing Dias to start his southward journey during the most tranquil season in the South Atlantic. As for Cão, the Crown had directed Dias to position stone pillars "at distances along the coast that seemed reasonable to him, [and] . . .

situated in noticeable locations."[3] However, his choices often lacked Cão's shrewd election of places with nautical significance.

On his journey beyond Cape Cross, Dias sidestepped a stupendous peaceful bay just seventy-three miles south of Cape Cross that would have served as an excellent place for a pillar. Protected by several miles of dunes, Walvis Bay hosted a wide variety of birds, aquatic mammals, and fish that flourished inside a shielded tidal lagoon.[4] However, having skirted this harbor, Dias sailed past miles of nearly vertical dunes before finding a partially shielded haven at a place he called Serra Parda, for the brownish color of the rocks along modern Conception Bay.[5] However, that anchorage was insecure, with too much of the bay exposed to the ocean. Dias had to sail another 171 miles alongside the same sandbanks before finding a suitable port.

The haven he chose consisted of three interconnected bays that a fifteenth-century observer deemed capable of holding a hundred ships of the era.[6] Despite its size, sailing between bays required considerable time tacking the sails back and forth to reach the other bays, leading Dias to name the place Anchorage of Turns (Angra das Voltas), known today as Lüderitz Bay.[7]

However, beyond its size, the Lüderitz harbor possessed minimal advantages. Although surrounded by wetlands that contained fresh water, the swampy ground limited the landing area. An unusually turbulent upwelling offshore separated the bay from usually nutrient-rich sections of the offshore current, leaving the bay with less variety than was found in other bays; here, only hakes, horse mackerel, and smaller fishes such as sardines and anchovies were found. With fewer fish came diminished numbers of birds (terns and plovers) and Cape fur seals.[8] Dias dropped off the second of the pillars at this bay, this time in a visible location.[9]

However, this stopover in Lüderitz Bay did not relieve the desolate shoreline as Dias and his crew continued down the coast. About a hundred miles south of Lüderitz, they encountered a series of rocks offshore and waters filled with fish.[10] As they proceeded, they named points along the way in a slapdash fashion, leading João de Barros to remark: "[Dias's crew] also named capes, coves, and the appearance of the land they discovered, either because of the day they arrived there or for any other reason."[11]

Finally, just beyond twenty-nine degrees south, Dias's ships located a modest rocky anchorage near today's small South African town of Holgat.[12] The first foreboding sign appeared as the boats prepared to leave the harbor. They attempted to depart for five days but were beaten back by freezing winds and the Benguela Current rushing northwestward.

On the sixth day, fortune appeared to smile on the fleet. Although still rustling in the opposite direction, the breezes were mild enough to allow the ships to proceed southward.

Barely beyond the port, the ships were trapped in mile after mile of wholly unexpected storms. Since no one had ever navigated this area, they were unprepared for the broad band of gale-force winds that stretched across the entire width of the Atlantic from West Africa to the east coast of the Americas. It is difficult to stand, let alone sail, in sustained surface winds of thirty-four to forty-seven knots (i.e., 39–54 miles or 63–87 kilometers per hour) while fighting the constant up-and-down motion of seas thirteen to twenty feet (four to six meters) high.[13]

The contrary directions of winds and currents only added difficulty to Dias's mission, which entailed sailing southward. Yet the storm's winds and currents constantly shoved the boats in the opposite direction—westwardly (away from Africa) and northwardly (back toward the equator). Only ruthless determination, the pilots' outstanding navigational skills, and the sailors' unceasing labors enabled the fleet to remain upright as they struggled for nearly two weeks under exhausting conditions.[14]

Suddenly, on the fourteenth day, the winds shifted. Instead of yanking them out into the Atlantic, fresh breezes began blowing eastward at seventeen to twenty-one knots, back toward the sought-after coast of Africa. At roughly the same time, an equally helpful current (0.9–1.0 knots) began to flow toward the northeast.[15] The boats had lucked out, allowing them to escape the deadly embrace of these ocean-wide gales and enter a relatively tranquil zone in the southern Atlantic.

As nighttime skies cleared, Dias checked his latitude by the Southern Cross constellation, estimating his position at thirty-five degrees south. At this latitude, the pilots and crew had managed to sail approximately three hundred nautical miles south of their last harbor under enormously challenging conditions.[16] Although Dias had no way of knowing it then, his escape during those hellish two weeks turned out to have favorably positioned his ships. At exactly thirty-five degrees south, he was only 8.4 miles below the southernmost point of the African continent.

Since Dias did not report his longitude, we cannot know how far west he might have sailed during the storms. However, the strongest winds blowing him back toward Africa lay between two hundred and five hundred nautical miles west of Cape Town.[17] After escaping the storms, a following wind favored the boats as they headed east, and the crews soon glimpsed Africa's southern coast.[18]

When they first glimpsed land, the crew was astounded. Instead of appearing directly east of the ships, as it had on all previous voyages south of the equator, the shoreline lay to the north. However, their excitement remained tinged with reservation. They just might have found yet another gulf. Two decades previously, ships had come across a similarly situated coastline. But the Guinea coast had stretched eastward for a thousand nautical miles before ending in the Bight of Biafra, where ships encountered an extensive landmass blocking their journey eastward. Thus, several thousand miles south, the crew remained unsure whether they had merely reached another lengthy gulf or the long-sought sea passage to the Indian Ocean, and the ships continued cautiously from bay to bay along the southernmost coast of Africa. Less than four hundred miles later, as the boats entered the much warmer waters of the Indian Ocean, it dawned on the crew with absolute certainty that no land would further block their progress, and they had indeed traversed the waterway between the Atlantic and Indian Oceans.

However, the exhausted crew first needed to anchor, rest, and resupply their ships; the southern African coast provided many such excellent opportunities. Notably, in this encounter, Dias initially approached a river with droves of cattle along the shore.[19] Given the large numbers of accompanying cattlemen, Dias unimaginatively named it the River of Herders (Rio dos Vaqueiros). Anxious to replenish the ships' water, he made minimal efforts to communicate with these locals taking their cattle to drink at a freshwater well near the shore. After making a half-hearted attempt at communication by throwing some goods on the beach, Dias charged ashore as if entitled to take the wellwater. His brusque arrogance and failure to negotiate permission angered the local herders, who began to throw stones at the Portuguese crew to drive them away. According to one account, Dias responded by killing one of them with a crossbow from no more than 320 meters away (just over a thousand feet).[20] Dias and his crew promptly left the River of Herders.

However, the three ships continued eastward along the coast without further incident, reaching a more extensive, sparsely inhabited bay that he named for the saint whose feast day it was, Saint Blaize (the iconoclastic Dutch renamed the bay after the abundance of mussels, hence today's Mossel Bay). Since the translators on board knew nothing of the local language, the ships acquired a supply of seafood either directly or through an exchange that concluded without any significant incident.

As the fleet proceeded cautiously eastward along the shore, they kept an eye out for the possibility of land straight ahead. Next, they passed a high peak they named Star's Hill, followed by an abundant fishing ground they called, again unsurprisingly, "Large Fishing" (now Gerricke Point). Then came Plettenberg Bay, the fires around Cape St. Francis labeled Burning Point, a second gulf populated by herders, and Rock Cape.[21] To the crew's surprise, the coast remained port side throughout this journey and eventually began trending increasingly northeastward.

Finally they reached a large bay (today's Port Elizabeth), where the coastline turned decidedly northward and sea temperature increased. These warmer waters further reinforced the impression that the ships had reached the long-sought maritime route around Africa. Then, anchoring off a group of small flat islands on the harbor's eastern edge, they found abundant marine life and skies filled with birds that were feasting on the giant schools of fish.[22]

Dias placed another pillar on one of the islands, after which the captains, pilots, and high-level crew beseeched him to halt the journey and return to Portugal. They were exhausted from nearly two weeks battling grueling South Atlantic storms. They had fought the formidable Agulhas Current and its retroflection as they sailed between bays along the South African coast. The crews were also concerned that the food supply would barely suffice for the return journey. Besides, they added, it had already become clear that they had reached the Indian Ocean.[23] Despite this vocal opposition, Dias pleaded for another two or three days, convincing the crew through threats and promises.

However, this truce with his crew lasted only three days at most. One hundred nautical miles north, the crew completely rebelled: No more. They had rounded the southern tip of the African continent, proved the existence of a sea linking the two oceans, and even started heading north into the Indian Ocean. Dias was forced to give in and planted his easternmost pillar on the banks of the Keiskamma River.

Although he resented their rebellion, the crew saved Dias's life and their own. As Vasco da Gama would discover more than a decade later, the next segment of the ocean was fraught with crushing difficulties that his fleet only barely managed to surmount. Had they not rebelled, Dias's already exhausted crew would have become just another casualty of the treacherous Agulhas, and the discovery of a maritime link between seas would have disappeared, like many others, into the depths of the Indian Ocean.[24]

Only on the return journey did Dias first spot the Cape of Good Hope, as westward currents carried him close to the cape. This pattern was later seeded by hundreds of boats returning from the Indian Ocean.

An enduring legend claims that Dias called the place the Cape of Storms, but that King John II insisted on naming it Good Hope because it marked the promising start of an entirely seaborne route to Asia.[25] Although Dias purportedly placed a stone pillar there, no trace has ever been found.[26] Continuing north between the excellent harbor of Cape Town on the Atlantic Ocean and Cão's Cape of the Cross, Dias enjoyed both a following wind and currents heading his intended northwestward, complemented by a strong sea and land breezes that eased a swift return.[27] Dias left his final pillar in a small, sheltered harbor now called Port Nolloth, just south of Lüderitz, another mediocre choice for a navigational marking.

Upon reaching Lüderitz, Dias expected to rejoin the nine men he had left behind on the supply ship to gather information about the residents while he moved on. However, when he returned nine months later, he found only two survivors, one of whom was the scribe for the voyage, who presumably created a written record of the contacts.[28] Illness caused only one of the seven casualties, "and the reason they gave for the [other] dead was because of the blacks of the land with whom they came to have contact, who killed them out of greed for some things that they traded."[29] In other words, six proved unable to carry out trade peaceably and were killed. Dias then transferred the remaining goods from the supply ship, which he then burned because its pilot was among the dead.

In his final stop during the return journey, Dias stopped at the Trade (Resgate) River and then at Mina to profit from his trip. In his instructions, the king explicitly ordered Dias not to seize captives or enslave people along his way.[30] Since the royal directives had only required exploratory voyaging, Dias's desire to return with a profit revealed his priorities, not those of the leaders who had organized and authorized the expedition.

The final twist in Dias's careless and indifferent behavior came from his actions regarding an unusual request from King John II for a novel approach to making initial contact with previously unknown peoples. The king's advisors selected four women from the Guinea coast (likely enslaved people), who were taken aboard Dias's ships, outfitted with fine clothes, and given silver, gold, and spices. Each woman would be left at an inhabited place along the route, where she was to show off her possessions, inquire whether similar objects could be found locally, inform the regional population of the glories of Portugal, and ask for the whereabouts of a possibly mythical Christian prince, Prester John.[31] According to the instructions he handed Dias, the king considered women to be more effective ambassadors and less likely to be attacked even if they encountered local strife.[32] Since no further information appears, it remains impossible to know whether the king acted on sexist stereotypes based on his own experience or information received from earlier contacts with more northern communities.

In principle, these women would be collected by either Dias's or a later fleet on a return voyage. However, one of the four women on Dias's expedition died en route, and two of the others were never found when Dias returned. Only the fourth, whom Dias had left near Port Elizabeth, was successfully retrieved. In a final cruel twist, when returning to the same islands where the crew had first rebelled, Dias abandoned one of his own sons into permanent exile.

His pilot's and crew's revolt demonstrated that Dias lacked the flexibility and leadership to compromise without conflict. Whether through a desire for the glory of being the first to reach India or mere arrogance, his narcissism left him unsympathetic to the exhaustion of the men serving under him. His further indifference to the proper way to introduce exchange with recently encountered people appeared in his clashes at the Bay of Cowherds, and he failed to instruct the crew in Lüderitz on appropriate protocols for initiating commerce. The latter omission led to the deaths of six men and perhaps the woman from Guinea as well. Finally, the unscheduled stops at the trading point and at Mina fort—as he said, "in order not to return empty-handed"—suggest an unwelcome degree of avarice.[33]

However, one of his navigational achievements endures. The most recent (2020) official U.S. guideline for sailing this route could have been written by Dias in 1486: "If eastbound from Algoa Bay, in favorable weather, Bird Island Passage is recommended, as it avoids the S[outh]W[est] current always running outside."[34] Dias had commenced his northward journey from farther inside the bay on an island called "The Cross." But he continued to the low-lying Bird Islands, recognized the swift-moving current to starboard, and proceeded along these islands (the Bird Island passage) into the heart of the southern Indian Ocean. On the next journey through this bay, da Gama's crew breathed sighs of relief as they emerged from a storm and found themselves within sight of the Bird Islands.

Thus, even though he had demonstrated outstanding nautical talent, Dias had not proven himself a great leader of men. Nor had he shown the necessary tact to negotiate for water or trade with newly contacted people. Hence, after his return to Lisbon, the king kept him busy for the next five years, working to improve ship design so that the next set of boats would be more stable and better able to weather the stormy waters of the southeastern Atlantic.[35] Although he may have chafed at his restriction to land, Dias remained marooned while others were sent to explore new directions and uncharted lands. Nor would he receive another significant command.[36] The leadership of the next fleet was given to da Gama, and that of the following flotilla to Pedro Álvarez Cabral. While Dias was included in this second large convoy, he had charge of only a single vessel belonging to a twenty-seven-boat squadron. Dias and his ship perished in a storm near the Cape of Good Hope, close to the place he had first glimpsed from the deck of a boat a dozen years before.

Dias's voyage proved the previously untested prospect of finding a waterway linking the Atlantic and Indian Oceans. Some hint remains that three voyages might have followed, inching partway up the coast of the Indian Ocean. However, eleven years would pass before the triumphant voyage connecting Lisbon and India by sea, led this time by a man whose leadership of his crew and contacts with natives were the opposite of Dias's: Vasco da Gama.

11

Crisscrossing the Atlantic

1497

The sources fall silent for the eleven years after Dias's return from beyond the Cape of Good Hope. The absence of reports of sea trials seems less like a temporary halt and more like a void left by the tragic 1755 disaster that cast a shadow over maritime history. During these "missing" eleven years, though, Portuguese navigators made two groundbreaking discoveries: one about the uniquely shifting paths of the mid-Atlantic currents and the other about the opposite directions of significant currents north and south of the equator.[1] These discoveries would revolutionize maritime navigation and exploration.

However, the sources leave only scattered hints of how sailors learned these techniques. Vasco da Gama's diary abruptly ceases upon his ships' departure from the Cape Verde Islands, leaving a gap in the record. The fleet "vanishes" while still in the Northern Hemisphere, at Port Praia, 14.9 degrees north; only to reappear (as if by magic) at 34.2 degrees south, in what is today the Western Cape province of South Africa. These ninety-three days, the most prolonged continuous period at sea, were shrouded in secrecy, with only isolated sentences hinting at the course they sailed or the challenges they overcame. Yet we can reconstruct how they might have navigated this feat by deciphering how the Atlantic currents flowed.

THE INTRICATE CROSS-EQUATORIAL DANCE

In the spring of 1497, four ships with crews of exceptional bravery and talent weighed anchor from the river near Bethlehem Tower, captained by Vasco da Gama, his brother Paulo, Nicolau Coelho, and Gonçalo Nunes. Their masts gradually disappeared over the horizon, bound for the open waters of the South Atlantic. Like many previous southward journeys, including Bartolomeu Dias's, they initially headed to the Cape Verde Islands. But the fleet headed in a distinctive direction when they left Cape Verde's São Miguel Island in August. Instead of proceeding southeast toward Africa, da Gama's ships sailed west, beyond the mid-Atlantic ridge where the Azores lay, for the first time in Portugal's more than sixty years of Atlantic voyaging. The crewmen, displaying remarkable calm and confidence, were about to embark on the first intentional use of the mid-Atlantic's crisscrossing equatorial currents, the most complicated in the world.

The ship pilots' apparent self-assurance suggests prior knowledge of the currents, though only indirect evidence remains. It is possible that the Portuguese made an effort to map these currents in secrecy; for example, the Crown made multiple payments to individuals for shipboard biscuits between 1488 and 1494 without noting anything further about the voyages for which they were intended.[2] However, we can likely pin a more specific date to at least one discovery. During negotiations with Castile in 1493, King João II asked only for exclusive travel rights one hundred leagues west of Cape Verde. However, the following year, he nearly quadrupled his request to 370 leagues beyond Cape Verde. This shows that sometime between 1493 and 1494, the Portuguese Crown learned precisely how far west a ship needed to sail to cross the Atlantic and reach the southward-turning currents. Since the Portuguese requested a distance east of the area claimed by Castile, they agreed to extend Portugal's exclusive navigational zone in the Treaty of Tordesillas, signed in 1494.[3] That they required a half dozen years to determine how far west they needed their ships to travel was a result of uniquely complicated currents.[4]

EQUATORIAL CURRENTS

Equatorial currents always travel westward. For example, the Canary Current joins the North Equatorial Current and flows westward in the North Atlantic.[5] Like such currents in other oceans, it always remains north of the geographic equator. However, unlike similarly named currents, the westward equatorial current never entirely crosses the Atlantic. Instead, about three-quarters of the way, it turns northwestward, heading straight to the Caribbean without ever coming anywhere near the coast of South America.

Equatorial countercurrents, by definition, always travel in the opposite direction (eastward), except in the Atlantic. The Guinea Current, which carried ships to the trade in gold and enslaved people, traveled mostly eastward, the expected path of the North Atlantic Equatorial Countercurrent. However, beginning roughly south of Africa's westernmost point, at Dakar, this equatorial countercurrent ceased to behave like all other countercurrents. First, it turned westward from the longitude of Dakar until south of the Cape Verde Islands. From that point, it became even more challenging. It continued southwest of these islands for sixty miles, then turned north, and then south.[6] At one point, the current even split in two, simultaneously moving due south and northwest at the identical place in the ocean. In its final unique turn, it joined the North Atlantic Equatorial Current and flowed into the Caribbean. The ships would have to cross another set of equatorial currents to travel southward.

No record reveals how and where da Gama's fleet crossed the disorganized North Equatorial Countercurrent; centuries later, British guides revealed that some ships would take advantage of the band of storms in the Intertropical Convergence Zone to head south. Regardless of the details, choosing to head south in the middle of the Atlantic, thousands of miles from land, demanded courage, conviction, and a keen understanding of ocean weather.

But da Gama's ships would have needed to traverse two more currents south of this chaotic countercurrent. Between three and two degrees north of the equator, his fleet would have encountered the first of three westward equatorial currents labeled "Southern."[7] South of this first narrow band lay a broader but slow-moving westward current, officially labeled the Central South Equatorial Current. However, the fleet could not follow either branch of the South Equatorial Current because they, too, never reached the coast of South America. Instead, those currents both joined the North Equatorial Current and Countercurrent to form a powerful, fast-moving, three-hundred-mile-wide Caribbean-bound current, a critical feature of the Atlantic.[8]

Only the Atlantic, among all the world's seas, moves warm water north across the equator. These four equatorial currents and countercurrents carry sun-heated water into the Northern Hemisphere, keeping the eastern United States and Western Europe warmer than if they were at the same latitude along the Pacific coast.[9]

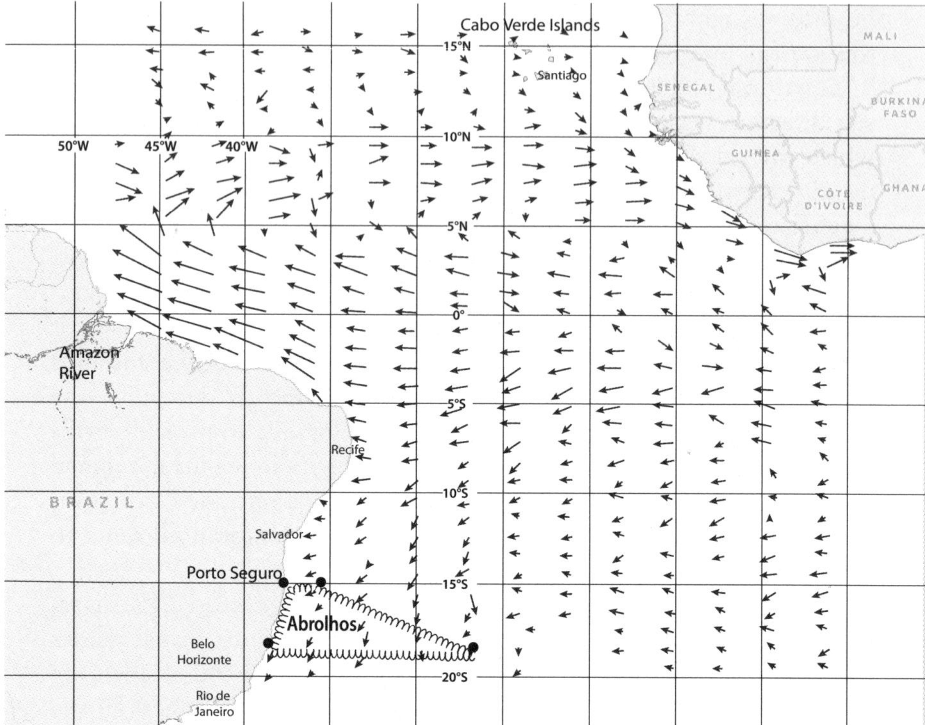

MAP 12. Equatorial currents that Vasco da Gama crossed to reach the Southern Hemisphere. © Esri, Tomtom, Garmin, Open Street Map Contributors and GIS User Community. Data from Polonsky et al., "Study of the Equatorial Atlantic Currents," 2017.

This rare diversion meant that Portuguese pilots needed to pay close attention to signs of a coming northwestward shift, lest they risk winding up in the Spanish-controlled Caribbean. While the Spanish ships intercepting the Portuguese boats treated them as spies, which they may have been, the pilots could just as readily miss the signs and unintentionally wind up in the Caribbean Sea.[10]

Besides shaping the climate in the North Atlantic, the Caribbean Current also prevented ships that had deliberately or accidentally crossed the Atlantic from returning to Europe. Its fast-moving, three-hundred-kilometer width kept vessels that might have arrived in the New World from reaching any possible return currents. They might have traveled north, to the Caribbean, or south, as we will see shortly.

In addition to these climate-forming and return-blocking attributes, the equatorial currents possessed a third exceptional characteristic.

Around fourteen degrees south, the southern branch of the South Equatorial Current split into opposite directions, one part turning southbound toward the freezing winds and steep waves of the Antarctic Circumpolar Current.[11] When da Gama's fleet joined the southward Brazil Current, they were the first in recorded history to use the opposing directions of winds and currents, or the Coriolis effect, to sail in separate hemispheres.

THE CORIOLIS EFFECT

Water and wind circulate in different directions north and south of the equator. In the Northern Hemisphere, giant currents circle clockwise from North America to Europe and back toward America; the pattern in the Southern Hemisphere is the opposite. From southern Africa, currents flow counterclockwise north to the equator, then west toward the Americas, turning southward with the Brazil Current toward Antarctica, before returning eastward across the Atlantic to South Africa. In an episode of *The Simpsons*, Bart goes to Australia, where he discovers that toilets flush in the opposite direction there, compared to the toilets back home in Springfield, USA. While the counterclockwise flush actually results from a different toilet design, the story offers a humorous mnemonic.

It appears likely that the Portuguese were aware of this phenomenon. The vast sums invested in outfitting the fleet and the crew's aplomb suggest confidence about finding a southbound western boundary current, possibly derived from unrecorded exploration during the eleven silent years. Thus, with its southward turn, the fleet followed the customary course of western boundary currents in the Southern Hemisphere, employing the as-yet-unnamed Coriolis effect to glide southward roughly in parallel with the coastlines of the modern states of Bahia and Espiritu Santo.[12]

BACK TOWARD AFRICA

As he followed this counterclockwise route, da Gama would need to know how far south to sail—since the winds and currents continued to the southernmost extent of what is now Argentina's Buenos Aires province—as well as where to begin the transition to head east to Africa. As he headed south, he encountered three possibilities: three counterclockwise eddies that would have facilitated a turn to the southeast, leading back across the South Atlantic to Africa. These eddies swirled around three underwater objects close to shore.[13] While this detail could easily have been lost to history, a momentary break in the journal's silence indicates which reef they would have used to make the turn:

> [August 22] Sailing around the sea, heading south by southwest, we found many birds flying back from the sea to the south, and a quarter southeast, we found many large green to blue-green birds [giant frigate birds]. When night came, they took off against the southwest very doggedly like birds do when heading towards land. And this same day, we saw a whale.[14]

These fish-eating birds indicated abundant edible sea life that the sailors on board would have welcomed. However, the presence of the whale provides the most unambiguous indication of the ship's approximate location, since whales only appear around one reef in the area.[15] Between sixteen and twenty degrees south, the third large area of reefs extends far out from the continental shelf. The reef the Portuguese named "Thistles" (Abrolhos) is created by floating red algae that clump together in ever-larger groups that float in the water. Each unique cell of red algae creates a hard shell of calcium carbonate. Eventually, the jagged clump of floating "thistles" becomes too heavy to be buoyed by undersea currents and falls to the ocean floor, where they pile into jagged and irregular formations, much like coral. This unique ocean ecosystem covers eight thousand square miles (20,904 km^2) and provides an abundance of food for whales that strain the ocean for food: the humpback, southern right, and Atlantic minke. Hence, we know that da Gama was in this region on August 22, 1497.[16]

Three years later, an unexpected storm blew the second India-seeking fleet westward at the northern edge of the Abrolhos Bank. As the storm cleared, Captain Alvarez Cabral became the first to see the coast of Brazil. After a week of exploring, Cabral weighed anchor at 16.43 degrees south. Had da Gama's ship passed sixty miles closer to the shore, his crew would also have seen land north of where they turned southeastward, which suggests that da Gama might have reached Brazil, had he encountered a similar storm.

At the southern end of the Thistle Bank, da Gama would have found a unique fishing zone full of giant green sea turtles and a multitude of unique tropical fish, including a blue-gray triggerfish that behaved like a piranha in the water, squealed like a pig when caught, and proved delicious when eaten.[17] Leaving Abrolhos, da Gama would have followed a slow eastward current along these fishing grounds, shadowing a chain of submerged volcanoes extending 621 nautical miles east of modern-day Vitória, the capital of the present-day Brazilian state of Espíritu Santo.

On the following two fleets bound for India, navigators caught sight of the volcanoes along what has become known as the Vitória-Trinidade Seamount Chain.[18] In 1501, during the second voyage to India, Captain João da Nova spotted a 570-foot (175-meter) rock rising out of the sea. This long-extinct volcano, later known as Martim Vaz island, marks the very end of the seamount, but its steep sides and rocky outcroppings are open to winds and currents, providing no place to land. The following year, a captain on the third voyage to India caught sight of another, taller—two thousand feet, or 620 meters—extinct volcano rising just before Martim Vaz. Later known as Trinidade, this enormous volcano boasted a harbor where his fleet anchored briefly, finding several streams to replenish their fresh water. Despite not seeing volcanoes, the routes of both successive voyages suggest that da Gama first followed the fish eastward along this route.

Having traveled between five hundred and six hundred nautical miles eastward to the end of the volcano chain, no one knew how far they were from the Cape of Good Hope.[19] Unbeknownst to them, easily 60 percent of the cross-Atlantic voyage remained.[20] Furthermore, they would need to travel south-southwest against a slow northwestward flow. This extended use of current sailing involved continually correcting course to the southeast for over two thousand miles. To keep heading toward the Cape of Good Hope, da Gama's fleet sailed along a single trajectory, or *rumo*, a term in fifteenth-century Portuguese that meant "direction or route." This imperfect initial solution arose from the lack of landmarks. On the open ocean, well out of sight of land, the principal source of direction appeared as lines on a compass. By the 1490s in Portugal, these compass lines had become known as *rumos*, indicating a straightforward trajectory. This journey would not take them along a single set of currents. Once past the southwest convergence zone, they would sail at right angles to the currents.[21]

However, unpredictable winds complicated the journey. Blowing erratically north one minute, south the next, and then suddenly westward, these winds dithered, as if unable to decide what direction to blow. However, at a very reasonable eleven to sixteen knots (thirteen to eighteen miles per hour, twenty to twenty-eight kilometers per hour), these unreliable winds blew sufficiently hard to help the ships move without threatening the lives of those on board. They only demanded unwavering attention to their ever-shifting direction, demanding that sailors adjust the sails to remain on a steady east-southeastward rhumb.

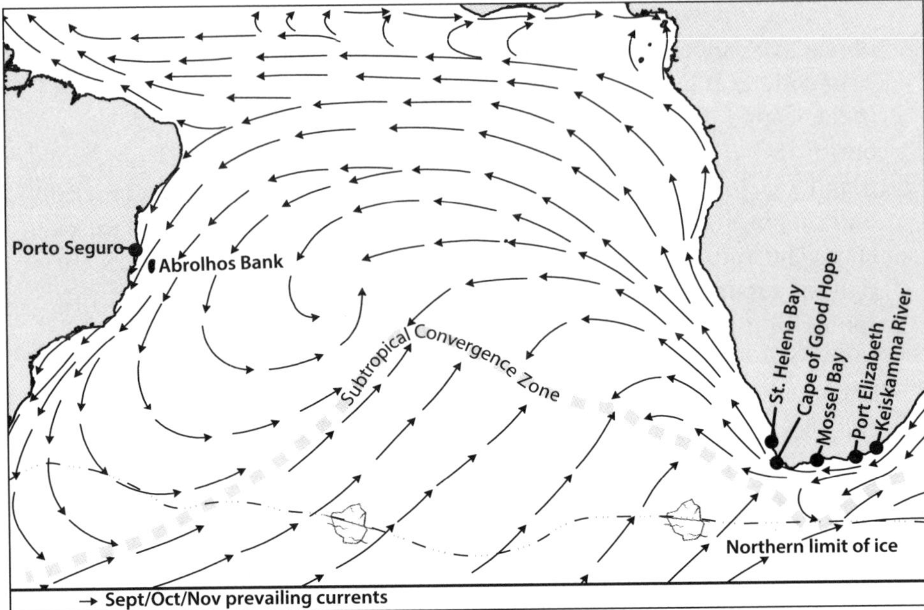

MAP 13. Transatlantic currents that da Gama's fleet encountered. © Esri, Tomtom, Garmin, Open Street Map Contributors and GIS User Community. Data from U.S. Hydrographic Office, *Pilot Charts South Atlantic*, 2009.

This final Atlantic crossing continued for forty-four days, covering nearly three thousand nautical miles at sea during which they were out of sight of land. Finally, on October 27, 1497, the report notes that they saw "whales, some [animals] that we call seals, and sea wolves [sea lions]." Five days later, "we found many signs of land, such as some dolphins born along the coast."[22] Here again, the diary identifies proximity to land by the appearance of recognizable sea animals. Unlike the ordinary seals with which Portuguese sailors had long been familiar, similar-looking sea lions reside only near the North and South Poles. Their small triangular ears on either side of their head and large side flippers distinguish them, and they indeed spend long periods on land. The hungry whales and dolphins the diary made note of were there to devour the vast quantities of fish that feed offshore from May until late November.

Only three days after the dolphin sighting, on Saturday, November 4, 1497, the report continues: "Before dawn, we found the sea depth a maximum of 120 fathoms [220 meters, 722 feet], and by 9 a.m., we had seen land." All of the crew, save for da Gama himself, gathered on deck dressed in party clothes, where they flew flags and fired cannon shots in celebration. The ships took a celebratory lap, sailing close to the shore before returning to sea.[23]

Despite the celebration, the Portuguese remained cautious. Even after three straight months at sea, they postponed landing until they had identified a safe harbor. Three days after spotting land, they came across a wide bay where they sent the pilot Pero D'Alenquer out in a skiff to test the safety of the sea bottom for anchoring and determine whether the harbor would provide sufficient shelter from the winds. The following day, November 8, 1497, the fleet anchored in Saint Helena Bay, eighty-one nautical miles (150 kilometers) north of Cape Town, where they "spent eight days cleaning the ships, fixing the sails, and taking on firewood." They resupplied their fresh water from the decent-sized Berg River in the southeast corner of the bay.[24]

As their repair and resupply mission ended, they began tentative contact with the local inhabitants, likely the Khoikhoi.[25] Despite their disappointment that the locals possessed none of the valuable spices or metals they sought, the crew found they could trade their copper coins for various distinctive local crafts such as seashell earrings, penis sheathes, and fox tails that the locals shook in front of faces.[26]

After the trading ended, a crew member, Fernam Velloso, wanted to visit the locals' homes to find out how they lived and what they ate. The natives invited him to eat but opposed his visiting their homes. Velloso belligerently pursued his goal, however, infuriating the natives, who picked up their small spears. Then he started running along the beach, shouting for help, chased by the natives. A small group on board the flagship abandoned their suppers and jumped into a small sailboat to rescue him. When his rescuers disembarked, they were as close to Velloso as the natives, who began tossing their limited-range spears, slightly injuring three or four men, including da Gama. Velloso succeeded in returning to the main ship in the sailboat, an ignominious end to an ill-considered effort at ethnographic investigation.[27]

Eschewing further contact with the natives, the crew remained in the cove for another four days as they prepared for the next leg of their journey, reaching Africa's southern coast. They left Saint Helena Bay on the morning of November 16, heading south. The southern coast of Africa is uniquely shaped: Saint Helena Bay protrudes into the Atlantic, the coast past Cape Town is nearly straight, and the Cape of Good

Hope dangles like a bird's feather, narrowing to a point at the tip and then arching gracefully upward. The journey to the end of the feather took only two days, but rounding Good Hope's dangling point took twice as long, due to strong south-southeast winds. Finally, on November 22, the winds shifted slightly and the fleet sailed past the cape, then anchored briefly off Seal Island, on the other side of the cape.

Lying just offshore, Seal Island is a small, rocky outcropping, now populated by smaller species of seal. But when da Gama's fleet arrived in early December 1497, the island was inhabited by giant elephant seals, named for the males' long, elephantine noses. In the words of the report, they were "as large as enormous bears, fearsome, with giant teeth . . . and no matter [with] how much force a spear was thrown, it never even injured them." The island also held female seals and their pups; the report noted that giant male seals "roared like lions, while the smallest [pups] bleated like goats."[28] Sharing the rock was a colony of penguins that "brayed like donkeys." This distinctive call has given the black-footed South African penguins a nickname: "jackass" penguins.[29] In the centuries since then, the elephant seals have migrated nearly a thousand nautical miles south to Marion Island and Prince Edward Island, near Antarctica. However, the black-footed penguins have remained, though they no longer stay on Seal Island in False Bay.

Three days after their stop at Seal Island, the fleet paused at Mossel Bay to dismantle their supply ship. Supply ships served an essential function: No vessel at the time could carry three months of rations, so to avoid dehydration, starvation, and scurvy a dedicated boat was built to serve as a floating depot. Once this depot emptied, sailors would stop mid-voyage to dismantle it, dividing its supplies and reassigning its crew among the remaining ships. On December 8, 1497, the boat's last pieces went into the remaining ships' holds, perhaps to be reassembled in Lisbon.

Ferdinand Magellan's voyage around the world underscores the wisdom of this strategy. Without an accompanying supply ship to keep the crew fed, Magellan lost two-thirds of his crew to scurvy while crossing the Pacific, though his voyage was only six days longer than da Gama's. By contrast, Dias's and da Gama's voyages saw far fewer losses, underscoring the importance of a supply ship carrying additional water and rations on long open-ocean voyages.

Perhaps even more noteworthy is that the cross-Atlantic route that da Gama and his pilots pioneered became the principal seaway for the next five hundred years. Ferdinand Magellan used his route, as did

Francis Drake, James Cook, and hundreds of lesser-known navigators. Even to this day, it remains the major Atlantic route for modern container ships and oil tankers. The man who led the expedition that invented today's sailing routes deserves more than a footnote in nautical history, as the fleet continued parallel to the South African coast.

12

Encounters Along the African Coast and in India

Contrary to popular belief, local pilots played a limited role in the da Gama fleet's journey to the Arabian Sea.[1] For most of the distance, da Gama and his crew navigated Antarctic storms, powerful contrary currents, and lethally slow high-pressure areas without any information on the conditions ahead. It was only during the final third of their northward journey that local pilots appeared and guided them through the long-traversed Arabian Sea. Not long after they arrived from Saint Helena Bay, their first challenges began along Africa's southern coast.

MUSIC, DANCING, AND TRADE WITH THE HERDERS OF MOSSEL BAY

Six days had passed after anchoring in Mossel Bay. As the crew repaired da Gama's flagship, about ninety natives emerged from behind the bushes, gathering on the narrow stretch of beach, gazing at the ship. Warned by Bartolomeu Dias's report of a prior armed attack, da Gama waited briefly before sending a small armed contingent toward the shore in a rowboat. Once close to the coast, the rowers tossed a handful of inexpensive items across the water onto the sand. To their surprise, the natives picked up the goods and began wading toward the rowboat, where they could see additional goods. Da Gama was astonished—this response was the opposite of what he had been expecting.[2]

Despite the absence of weapons or any threatening moves by these people, da Gama remained cautious. The natives had initially stood on a narrow strip of sand where the dense, compact local shrubs called *fynbos* came dangerously close to the shore, potentially allowing hostile armed individuals to hide. Da Gama wanted to continue this friendly exchange, but in a more open area, so he signaled the rowers to move to a nearby part of the beach and the natives followed. Most of the Portuguese then cautiously disembarked, leaving only a handful of armed men on the ships. However, these precautions soon proved unnecessary, as a humdrum exchange proceeded uneventfully, with the natives trading the ivory bracelets they wore for small items, such as red hats. The following day, a more considerable troop of natives herded a dozen cattle and four or five sheep to the shore. They also brought their instruments to play music and dance. The Portuguese sailors then picked up their own instruments and joined the native musicians and dancers. At the end of the dance, the Portuguese traded three bracelets for a young boy to work on the ship—the only episode of enslavement during the entire journey. They then shared a very fat and delicious ox provided by their hosts, which they deemed "as tasty as any in Portugal." The following day, women and children joined the gathering, trade continued, and both groups played their instruments and danced together before the Portuguese departed.[3]

Despite the auspicious beginning, peaceable interactions did not continue. The next day, da Gama sent ashore a lone man who had traded extensively in the Congo to obtain another ox. However, instead of trading, the natives took Martim Afonso by hand to the same well that had led to such grief for Dias (see chapter 10). Afonso later said he thought the natives were asking why the Portuguese wanted the water from this well. We do not know how Afonso answered, only that in response, the natives drove their cattle into the safety of the bush.[4] The crew on shore went over to rescue Afonso, who was then running away from the fleeing natives. As a precaution, da Gama ordered other crew members ashore to carry spears and wear defensive armor. This show of force failed to have its intended effect. The natives only retreated, forming a group a short distance away. Da Gama ordered everyone back on the ships and fired two cannon shots into the water. The unfamiliar boom and loud splash terrified the natives, who fled from the shore, leaving clothing and weapons behind in their haste. Upon reaching the bush where their cattle were hidden, they united, drove their animals away, and vanished over a hill.[5]

When the natives remained out of sight the following day, the crew nervously approached the infamous well, where they refilled their water barrels without incident. They then removed a carved Portuguese limestone pillar from the ship's hold and placed it upright near the shore. Next to it, they erected a giant cross from a broken mizzenmast. The following day, December 7, just as they raised their sails and began to leave the harbor, the natives reappeared and started to take down both. The natives may have dismantled the two pillars to remove all traces of the foreigners in retribution for the violation of the well or even to acquire the raw materials, since the area had few large trees and no limestone. However, the Portuguese onlookers would have understood this as violating a quasi-religious symbol, rather than as harvesting raw materials.[6]

From these interactions, da Gama concluded that the hostile community that Dias encountered no longer resided beside Mossel Bay. But their insistence on interrogating Martim Afonso suggests that the people regarded the well as possessing some extraordinary power, either dangerous or sacred, and perhaps requiring a ritual approach. The preoccupation with understanding the foreigners' intentions regarding the well suggests that the group was most likely a closely related, if not the same, community. Once the crew had refilled the water supplies, finished repairs, and readied the fleet to sail again, da Gama headed, like Dias, toward the Indian Ocean. This time, however, the Portuguese fleet would encounter weather more typical of the coast.

THE SOUTH AFRICAN COAST

In its simplest terms, Africa's southern coast overlooks an ocean ruled by equally matched extremes. One of the world's hottest currents flows from the warm Indian Ocean, while icy Antarctic waters race up from the south. These winds and currents do not mix; they merely push each other, like children roughhousing on a playground, where who is pushing back more at any moment depends on which one is temporarily stronger. When the Antarctic storms arrive, deep gray clouds materialize on a distant horizon and icy Antarctic winds rapidly reach the shore, raindrops like long sharp needles plunge into the skin, and the air is so thick with rain that the long-suffering sailor can barely see ten feet away. Here, off the coast of South Africa, the Antarctic can conjure up such storms in an instant.

The unique geography of southern Africa further complicates the unpredictable weather. A long plateau mirrors the coast around nearly the entire country of South Africa. The air above the plateau heats up,

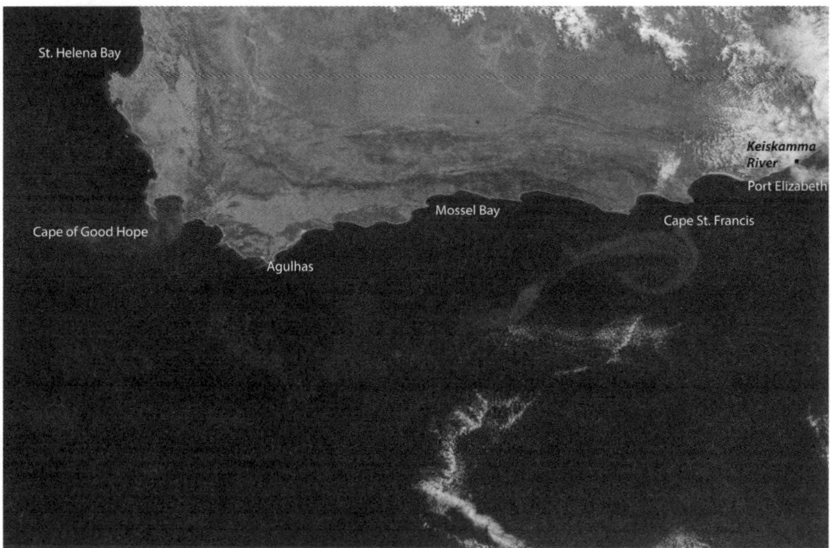

MAP 14. The South African coast. NASA.

especially during the summer (in December). As it blows toward the sea from the plateau, this hot, dry air—known locally as a "berg" wind—continues either for several hours or intermittently over several days.[7] Like the hot winds of North Africa, it blows at near gale-force strength—thirty-two to thirty-eight miles (or fifty to sixty-one kilometers) per hour—raising temperatures to above 40°C (104°F) and reducing humidity to less than 20 percent.

When the overheated air ceases to blow, cooler winds move in from the ocean, carrying precipitation ranging from fog or drizzle to intense storms. These storms or fogs become trapped between the plateau and the sea, moving eastward along this trough at nineteen to thirty-seven miles (or thirty to sixty kilometers) per hour before dissipating near Durban. Leaving Mossel harbor well supplied, the fleet started peacefully, with no hint of what was about to befall them.

THE TREACHEROUS AGULHAS

Calm winds slowed their passage on the first day out of Mossel Bay as they managed only thirty miles. The next four days passed uneventfully, but on the fifth the ocean had its revenge. Winds pounded the fleet as it descended into a nightmare of bone-chilling Antarctic cold. The poet

Luís Vaz de Camões, who sailed through these waters in the sixteenth century, described how a "thick black cloud rains down arrows and stones. . . . The black sea roars and lashes out from afar, as if throwing itself in vain against some high steep rock."[8]

Aboard da Gama's ships, powerful icy rain pelted the crew without ceasing. Gale-force winds whipped through the sails, which the crew had to lower to slow the boats and keep them upright. Their ships plunged abruptly down steep waves only to face angry towers of steel-blue and gray clouds racing toward them during their unceasing struggle to remain afloat. Amid this chaos, da Gama's crew also lost sight of the fleet's second ship, commanded by Nicolau Coelho, for much of the day. At sunset that first night, they managed to make out Coelho's vessel from atop the crow's nest. Since it appeared to be following them, they fired the cannons and lit a fire to signal to Coelho that they had seen him. But they quickly realized that any response he might have attempted would have been lost to the winds behind him.

The fleet only narrowly escaped a deadly threat sometimes created by the storms off Africa's southern coast.[9] When the Indian and Atlantic Ocean currents meet, small waves become trapped above large waves, forming a giant single lethal wave that sends vessels straight to the bottom of the sea without any hope of survival.[10] Da Gama managed to avoid these rogue waves, but these seas soon became a graveyard for scores of ships and hundreds of humans over the following centuries.[11] Such storms do not release their prisoners easily. But after three days, the pummeling winds and rains ceased. Early on the morning of December 15, the crew awoke to calmer weather. Having ridden out the storm, they could breathe a momentary sigh of relief.

As the sun rose, they began to take stock of their surroundings and soon recognized their location from Dias's description. The three ships had anchored on the eastern edge of today's Alagoa Bay, opposite modern Port Elizabeth near the eastern edge of southern Africa. Off the port side of the ship, they caught sight of a long flat rock jutting into the ocean and covered with unusual feathered animals. In the skies above, gracefully arched white-winged birds with dramatic black fringes wafted slowly in the gentle breezes. In an instant, these wings bent backward as the birds transformed into miniature white rockets, vertically dive-bombing into the ocean at twenty-five to seventy-five miles (40–120 kilometers) per hour. These missile-shaped fishers, now called Cape gannets (*Morus capensis*), plunge thirty-three feet (ten meters) into the water to target schools of fish, snapping up their meal before bobbing up on the surface, where they appear to be no more menacing than bathtub ducks, their momentary resemblance to deadly predators vanished. To the sailors, extensive flocks of diving seabirds could only mean an abundance of fish. Dias had named the Cape gannets' home and two smaller nearby rocks the Flat Islands (Ilhas Chãos). Later, the largest of these three "level islands" would be renamed for the avian residents and frequent visitors, becoming known simply as Bird Island.

Looking back across the bay, the crews of da Gama's ships saw the top of a distinctive cone-shaped hill twenty-six nautical miles away. There Dias had placed a wooden cross on that hill and christened the place Cross Island (Ilha Cruz), a name it bears to this day, only now with a French spelling, muting the memory of its origin. Having reached this landmark, the crew congratulated themselves on their navigational success. After all, by the time they had reached the Flat Islands, da Gama's fleet had effectively sailed along the entire southern coast of the African continent—a reported sixty leagues since Mossel Bay, and another sixty since the Cape of Good Hope. In modern terms, they had sailed 425 nautical miles around Africa. Although their distance measurements were slightly off, there is no mistaking their contentment.

Setting out north from Alagoa Bay the next day, they sailed northeast, past the final stone pillar that Dias had placed on a headland now known as Kwaaihoek, or "sweet water" in the local Khoikhoi language. Just beyond the pillar was a small freshwater river (the Keiskamma) where Dias's crew had rebelled and insisted on returning to Portugal.[12] Da Gama's fleet continued past the river that day without hesitation, remaining close to shore as they mapped the coast. They noted with relief that the land was fertile, well populated, and home to large herds of cattle and tall trees suitable for masts. However, this easy northeastward journey lasted only until the following evening.

On that Sunday, only thirty-six hours after they had departed from the calm of Alagoa Bay, powerful winds from the east swooped down across the boats, seemingly out of nowhere. To avoid being slammed against the coast, the crew immediately headed out to sea, where they had to tack back and forth against a sleek and powerful southbound current. After two days, the winds let up and shifted in the opposite direction, making it safe to approach the coast.[13] Toward sunset on the nineteenth, they anchored near the shore, where they spent the night. The following morning, they landed to find out how far they had traveled—discovering to their disappointment that they had managed to advance only sixty leagues, at most two hundred miles, over three days—and rightly blamed the current for their slow progress.[14]

Their actual stopping point has been the subject of some contention, but from a nautical perspective, there are only a few plausible candidates. The only safe anchorage along this entire stretch of the coast north of East London (beyond Dias's pillar) was today's Durban.[15] Furthermore, winds along this stretch of ocean change abruptly between warm air blowing from the east and cold air blowing from the west, a pattern that remains poorly understood.

Despite the crews' frustration with the slow northerly progress, they set out confidently the next morning, only to face a far more menacing force. The current they had been struggling unknowingly against on the earlier two days made up the outer edges of the Agulhas Current.[16] Its main body remained a considerable distance to the east, between Bird Island and Durban. However, once they passed Durban, they slammed into the main channel of the current, which resembles an octopus: stealthy, swift, and deceptively concealed deep in the ocean. It wraps its tentacles around the entire southern edge of the African continent, occasionally dipping a long, slender tip far out into the South Atlantic. Its arms furl and unfurl, trapping and erratically releasing giant circles of hot Indian Ocean water to float among its many other arms.[17] It moves seventy million cubic meters of water every second, traveling deeper and seventeen times more powerfully than its better-known counterpart, the Gulf Stream.[18] North of Durban, where da Gama recommenced his northward journey, it even outpaces that other current, shooting southward at an average of 5.8 miles (9.3 kilometers or five knots) per hour, making it the swiftest in any ocean.[19] Although only sixty miles (one hundred kilometers) wide, this "octopus" comes within twenty kilometers (eleven nautical miles) of the shore north of Durban. Da Gama and company were sailing in a narrow waterway beside a swift, powerful current flowing in the opposite direction.[20]

Upon reaching this narrow channel, the author of the report became apprehensive. The word for "fright" and an invocation of God appeared, followed by a reference to the Agulhas Current: "We feared greatly that the currents would not allow us to travel towards what we sought. And from that day forward, God wished that we traveled forward and not in reverse. Please God that it will be like this always."[21] The report thus invoked a supernatural power for the first time in the journey.

For the next five days, da Gama confronted one of his expedition's most difficult yet easily overlooked navigational challenges. This blind spot likely stems from the area's appearance on ordinary maps. Such images show deceptively open waters, suggesting a smooth journey to the southern tip of Madagascar. However, judging the realities of this passage by these renderings is like using a map of political boundaries to cross the Alps or navigate the Straits of Gibraltar. Such perspectives both flatten all heights and erase all dynamic movement. In this next critical section of the journey, da Gama faced an unannounced and unforgiving force of nature: the swift and opposite-flowing Agulhas Current with its almost chaotic retroflection.[22] As he headed beyond Dias's pillar, this current slowed their northward progress from a distance, unbeknownst to anyone on board. North of Durban, it was about to attempt to strangle him.

This five-day-long test of the crewmen and their ships ended on Christmas Day 1497, as the space between the current and the coast widened slightly and they were able to anchor safely near the shore, where they surveyed the ships and made emergency repairs. When sailing northward against the Agulhas, the lead ship had developed a lengthy crack across the top of the mainmast. The split had nearly severed the crow's nest, which the crew had addressed by using ropes to keep the top of the mast from shifting. Yet, as they went about more permanent fixes, the extent of their predicament became obvious. To their great disappointment, they had advanced only ten leagues over those five days. Later reports have da Gama naming the area Natal (meaning Christmas Day), but the actual report pays far more attention to the urgent need for repairs. Exactly where they stopped for emergency repairs remains somewhat unclear, the most likely place being at the heavily forested mouth of the Nenga River (−31.985) in an area that offered fresh water and partial shelter.[23]

Continuing north from this temporary stopping point, three days later they reached the excellent fishing grounds just outside of today's Durban, which they called Fishing Point, where they passed the day catching tuna, swordfish, mahi-mahi, and king mackerel, along with other forms of seafood.[24] From here, they continued north for another two hundred miles until they reached the originating region of the Agulhas Current, where they would attempt a second, previously unnavigated section of the southern Indian Ocean.[25]

On December 28, da Gama's fleet left the fishing grounds behind, unaware of the challenges ahead. Just over fifty nautical miles north, they bypassed a harbor with a decent sandy bottom but heavy swells. Little did they know that this seemingly harmless location (St. Lucia) held the last fresh water they would find for the next two weeks.

FROM THE MOZAMBIQUE CAROUSEL TO THE INDIAN OCEAN

A pivotal point occurred in the unexplored southern Mozambique Channel, where the crew had to ration water for the first time. Despite these challenges, they persevered, continuing north until finding a harbor with the large timbers needed to repair the damage from the Agulhas winds. During this month, and for the first time since their departure from Lisbon eleven months before, some crew members developed scurvy, having refused to eat the native fruits that could have prevented it. Setting off in the repaired vessels, they reached an island fifteen degrees south of the equator, where they first encountered ships that used monsoons to traverse the sea to India. Following a failed attempt to hire a local pilot, they succeeded in Malindi and crossed to India uneventfully.

THE LOSS OF FRESH WATER AND THE APPEARANCE OF AN UNCHARTED ISLAND

Just off the eastern side of Africa, north of the Agulhas Current, lay the island of Madagascar. It was a unique puzzle piece that did not fit anywhere along the coast. Eighty-eight million years ago, it had broken off and drifted southward from what became India. But this oddly shaped island parallel to the African coast created unique weather conditions, introducing a surprising and challenging twist in da Gama's voyage. These weather conditions, a result of the island's unique position and shape, added a layer of complexity to the already challenging journey.[26]

As the fleet sailed between Madagascar and the African coast, a wide, slowly turning carousel of winds formed at the top of the channel every fifty days and drifted southward. This chain of rotating high-pressure systems, at times nearly filling the sea separating Africa from Madagascar, was a sight to behold. Unaware of the island's presence to the east and its effect on the local weather, the Portuguese pilots were taken aback by the sudden and dramatic change in sailing conditions, including the shortage of the basis of all life: water.[27]

Having failed to anticipate a slow northward journey, they needed to resupply the ship with extra water. When crossing the Atlantic, they had a supply ship holding enough for three months. But having careened unexpectedly into the Mozambique Channel, they were caught unprepared by high-pressure systems that deprived them of rainwater.[28] With clear skies above, slow winds at sea, and fresh water concealed from view by the sandy coast, the crew experienced unprecedented hardship. They remained entirely at sea for the next two weeks, unable to find additional water. Near the end, they were forced to cook with ocean water and drink only half a liter of fresh water a day. Then, on January 10, 1498, they spied a wide sheltered bay with abundant streams. They had yet to learn how their presence would be received as they anchored offshore.[29]

THE GOOD PEOPLE

The following morning, early on January 11, a small party of crewmen set out for the coast. From a distance, they spied a small group they described as healthy, large-boned men and women accompanied by their leader. Since the group seemed more curious than threatening, da Gama selected one of his crew members to land and try to negotiate for some much-needed water. For this role, he chose Martim Afonso, who had lived in the Congo for a considerable time and had a working knowledge of Kongo, a Southern Bantu language.[30] The local leader welcomed Afonso using a different Southern Bantu language, but apparently with enough similar words that the two could carry on a basic conversation. During this exchange, the leader declared that his community would willingly supply the desperate Portuguese crew with anything they needed. Afonso returned to the boats bearing this welcome news, so da Gama gave him a jacket, red pants, a hood, and a bracelet to give to the leader. Martim Afonso returned to land with the gifts and spent the following night ashore at the leader's home.

The next morning, more evidence of local hospitality followed. After donning the clothing da Gama had given him, the leader reciprocated with good food for the crew, including a local porridge and a delicious hen—most welcome for a crew that had not eaten chicken in a long time. Soon, the settlement was revealed as far more extensive than the welcoming party. That next evening and the following morning, over two hundred men and women came out to see the Portuguese sailors and their boats. Many locals began ferrying fresh water from rivers and lagoons behind the shore. From these well-coordinated forays, the Portuguese concluded that the society was highly organized.[31]

In fact, the Portuguese had just encountered the Tsonga Tembé kingdom, established around Kosi Bay in 1300. Its subjects spoke Xitsonga, a version of Southern Bantu. They grew cassava, millet, and sorghum, trapped fish, and raised hens and cattle. The Tsonga organized hierarchical societies led by an elder whose status the Portuguese often later likened to that of a king. Over the next five days, the Tsonga and Portuguese continued trading. Of all the goods the Portuguese crew possessed, the Tsonga most prized their lightweight linen shirts, seeking to acquire as many of these vestments as the sailors were willing to part with.

Then the ships departed suddenly, despite the relaxed, ongoing exchange and continuing water resupply. During the five days in Inhambane harbor, the offshore winds had remained still. On the sixth day, fresh winds materialized unexpectedly, making it urgent to resume the trip immediately. Even though they had failed to resupply their entire store of drinking water, the ships returned to sea.[32] Because of their hospitality and generous donations of food and water, da Gama bestowed the designation Land of Good People. Likewise, he bestowed the name Copper River on the small entrance to the bay, after the copper bracelets and ornaments worn by the Tsonga.[33]

THE GOOD SIGNS

Having left the Land of Good People, the fleet traveled steadily north against successive carousels of high pressure. Although crossing the broadest continental shelf along the East African coast, the crews still had to dodge several extensive offshore reefs. Along the way, da Gama bypassed several mangrove ports, seeking a wooded region with enough timber to repair his boats. After eight days at sea, his ship reached a heavily forested harbor, including trees with broad trunks, some between twenty and sixty feet tall. And despite never having seen ships the shape and size of da Gama's, the natives welcomed the visitors.[34]

The surviving account furnishes few details about these people's language, customs, or rituals, making definitive identification impossible. It only notes that the men wore limited clothing, consisting of cotton cloth coverings, which they bought from a Muslim town in the interior. However, the women wore three lip piercings, which were traditional in many Southern Bantu communities, suggesting that they perhaps belonged to the Chuabo. Although lacking the prosperity of the Tsonga, the Chuabo proved equally generous to strangers. The author of the account wrote, "They brought what they had to our ships in canoes, and we also went with them to their village to bring [fresh] water."[35]

Two or three days after the Portuguese anchored in this harbor, two arrogant and expensively dressed men from the interior appeared: one wearing a cap with silk thread and the other a green satin hood. Both disdained any Portuguese gifts. Furthermore, these traveling merchants assumed that visitors would have similar desires for trade goods as the natives. Hence, they brought an ample supply of the simple white cloth the village men wore to sell. Having set up a temporary shelter on the land parallel to the boats, they began to trade their cloth. After a week, they retreated to the interior from whence they had come.[36]

Following the merchants' departure, the crewmen took advantage of the safe harbor and friendly natives to carry out the significant repairs the ships needed. For over a month, they made repairs, cleaned the hulls of all three ships, and prepared for the journey ahead, and at last they repaired the crack in the mainmast that had occurred weeks before when battling the Agulhas. However, during this month-long maintenance, some men developed scurvy for the first time in the nine months of the journey.[37]

Scurvy results solely from the lack of a single nutrient, vitamin C. The human body uses up its entire reserve of vitamin C within one to three months. After that, it begins to manifest the painful signs of the disease. Scurvy was somewhat common among sailors, who might go months or years without access to fresh fruit, but from the records we can tell that only a few of the crew developed the illness. It raises a simple question: Where did da Gama's crew get their vitamin C supply, even as the fruit stores from their supply ship ran low?

The account notes that the land produced abundant wild fruits that the natives enjoyed, wholly new to the Portuguese: small golden apples called Kei apples (*Dovyalis afra*) and shiny red Natal plums (*Carissa macrocarpa*), each holding more vitamin C than oranges; while the wild marula (*Sclerocarya birrea*) fruit's thick yellow peel and translucent white interior had more vitamin C than oranges, grapefruits, or lemons. A single marula fruit supplied double the daily adult requirement for vitamin C. Both Natal plums and Kei apples had a slightly tart taste, and the marula possessed a distinctive sweet-and-sour flavor. Hence, the crew members afflicted with scurvy had likely avoided the tart or sweet-and-sour fruits consumed by the natives and most of their fellow crewmates. Although the cause of scurvy would not be discovered for another 250 years, their willingness to eat the food enjoyed by a healthy local community saved most of the crew from the pains of scurvy.[38]

Before the arrogant merchants' departure a month earlier, a young man in their entourage had revealed that, in his faraway homeland, he had seen ships like their own. The crew rejoiced at this information because, in the words of the journal, "it appeared that we were heading to the place we sought."[39] As they prepared to leave on Saturday, February 24, da Gama placed a pillar along the coast and christened the river

"Good Signs" (Bons Sinais) for the welcome news they had received from the youthful visitor from a faraway land.[40]

SAILING NORTHWARD TO FIND A PILOT TO INDIA

Over the next week, the journey became more accessible and enjoyable. With fair winds and calm waters, they spent their days sailing, then anchored and rested at night. They even enjoyed sufficient leisure to remark on the beautiful view of a coastline and three small islands, two of them richly forested.[41]

After this week of ease, they glimpsed a vast expanse of land surrounding a large island and cautiously waited to approach the area during daylight. This decision proved wise, since the navigator of the lead ship under Nicolau Coelho miscalculated the tides and ran aground on a sandbar during low tide. Fortunately, he could signal Vasco and Paulo da Gama on the boats behind him to hold back. Coelho waited until high tide to free the ship and led the other two ships across the bar and into the waters around the Island of Mozambique. After that initial mishap, the boats anchored in the lagoon that served as the harbor for Mozambique, fifteen degrees south of the equator.[42]

As before, the newcomers soon drew the interest of local merchants—just as in much of the world today street hawkers race to the sides of trains or buses pulling into a station, offering food to those on board. In Mozambique harbor, food hawkers appeared alongside the boats, offering *anafes*, a type of celery often desired by sailors on large ships arriving from the north. Portuguese sailors were not as anxious for these snacks, so the men aboard the tiny boats retreated in disgust.[43] Unsure about what kind of welcome they could expect going forward, the Portuguese captains pulled back to plan.

Coelho and the two da Gama brothers then conferred about their next move. They needed to gather intelligence about the island and its people and decided that precautions were necessary. Initially, they had moored in the island's principal harbor alongside scores of other watercraft. Still uncertain about their reception, the captains decided to anchor outside the main harbor, where they could head straight back to sea should hostilities break out.

Da Gama dispatched Coelho to take soundings and determine the best place to station the ships. However, bad luck followed him again. Although he managed to avoid getting stuck on another sandbar, he broke his rudder as he rounded the island searching for a secure location.

FIGURE 1. Eighteenth-century French engraving of the sandbanks in Mozambique Harbor that detained Nicolau Coelho's ship.

Eventually, he did locate a spot where the winds could whisk them straight out to sea and where they would be at least two crossbow shots away from land. Given the extraordinary skills of Coelho's pilot, this day stands out as Coelho's single worst day of the entire journey. Nonetheless, he pushed forward and discovered a settled, hierarchical, highly developed seagoing culture.

The fleet's visit to the Island of Mozambique proved equally unexpected. They found four boats in the harbor laden with cloves, pepper, ginger, gold, silver, rubies, and pearls. But these wholesale merchants sold their goods by quantity, somewhat to the surprise of the Portuguese, who had not seen these goods traded in such large quantities before.[44] They also had their first astonished glimpse of coconut palms with fruits much larger than the date palms they were familiar with.

Portuguese navigators had found a maritime environment inhospitable to ongoing coastal communication during their previous half-century of voyaging along the western and southeastern African coasts. Some potential connections held only contrary or one-way currents, as off the coast of Bojador. Other routes were life-threatening even in sturdily built wooden ships, including the passage around the Cape of Good Hope. All of this discouraged sailing for Indigenous cultures. However, from Mozambique northward, ocean currents and winds blew with sufficient speed and predictability that towns could grow along the coast and create ongoing relationships linked by the sea.

Shipbuilding for this region placed relatively few demands on the artisans. Boats with open decks could sail between towns without needing covered compartments for protection from the elements. They did not require deep holds for the supplies necessary for months at sea, nor did they require high sides for long trips on heavy seas. These regional boats employed a single triangular sail fashioned out of palm mats and had a rudder attached to the stern. The sides of these regional boats were tied tightly with string made from coconut fiber, giving rise to the popular, albeit slightly misleading, name of "sewn boats."[45] When da Gama arrived, the captains of these boats had already adopted Genoese compasses and maps.[46] For the Portuguese, this opened the door to hiring a local navigator, which promised to both speed up the trip and teach them about the local sea routes.

Doing so, however, meant pleasing the local elites, a process that proved more problematic than their entrance into the harbor. The well-dressed merchant community wore multicolored cotton and linen garments and caps with silver and gold threads, which meant they could not be so easily dazzled by da Gama's wares. The governing sultan visited the Portuguese boats repeatedly, but these exchanges were unsatisfactory. The sultan rejected da Gama's gifts of hats, corals, and short capes with hoods and demanded scarlet, which da Gama did not have. Finally, da Gama sweetened the sultan's disposition with a dish of figs and preserved fruit; after that, he agreed to let da Gama hire two willing pilots for the next stage of the journey.[47] Da Gama paid each pilot nearly thirty grams (about an ounce) of gold, plus two capes, and the ships began heading north.[48] However, neither they nor their pilots were aware of the challenges the weather would present for the remainder of March.

On the first day out, they had barely cleared Mozambique harbor when they decided to stop at an offshore island and hear mass before continuing the journey. However, one of the local pilots decided to abscond and fled into the island's interior. Da Gama and Coelho promptly set out to recapture the errant pilot, who had gathered a small band of men armed with longbows. Da Gama and Coelho quickly recaptured the pilot and fired their weapons for the first time since leaving Cape Verde, forcing the remaining boats to flee. Although this initial effort to acquire a local pilot could hardly be termed successful, it marked a new phase in the journey around Africa.[49] Likewise, the need for fresh water resurfaced again just north of Mozambique. As earlier on the trip, the absence of wind prevented the fleet from progressing. However, unlike their previous experience, the area north of Mozambique had a sizable coastal shelf, which made anchoring easy. Despite their safety, the fleet began to run out of fresh water within a week. Anchoring along the way, they engaged in contentious, but ultimately successful, negotiations for fresh water.

A WELCOME IN MALINDI

Soon the Portuguese encountered boats near a village along the coast. "On the same day [April 14], at sunset, we launched an anchor in front of Malindi, 30 leagues from Mombasa," where both pilots from Mozambique had jumped overboard and fled.[50] They approached the harbor because they needed someone to guide them across the Arabian Sea. They knew of the regular monsoon routes to India, which had existed for over a thousand years. Still, having never used them, they wanted a local pilot to ensure their safe and expeditious passage.[51] As in Mozambique, this would involve the challenging process of first contact, as well as navigating a highly unfamiliar culture.

Malindi (known as Melinde in antiquity) is a town on the Kenyan coast sixty-five nautical miles northeast of Mombasa. The welcome in Malindi could not have been more different from the greeting in Mombasa. The Portuguese text referred to the local ruler as a king—a term of respect reserved in fifteenth-century texts for friendly heads of state, hence the king of Congo, the king of the Tsonga, and the king of Malindi.

The king sent the captain six rams, many cloves, cumin, ginger, nutmeg, and pepper, and offered to meet him on the water in the harbor. A little later, each arrived at the meeting place in a small boat. After they had talked for three hours, da Gama turned over unharmed all the crew he had captured at sea. At the end of the meeting, they exchanged hostages, as was customary during peace negotiations. For the next nine days, the crew enjoyed regular entertainment, including music provided by a combination of trumpets and ivory tusks, which were as tall as a man but played as a wind instrument. By the time the crew had finished resupplying the ships for the short trip across the Arabian Sea, the king of Malindi had given them a pilot as a gesture of good faith. India seemed in reach.

ARRIVING IN INDIA

With the southwest monsoon in full swing, da Gama left Malindi on April 26, spending twenty-three days at sea with a continual southwest wind. By May 18, they were once again in sight of land and two days later had anchored along the coast of Calicut, which the king of Malindi had assured them would be an excellent place to acquire the spices and secure trade agreements. The journey across these seas had taken only one-quarter of their time crossing the Atlantic.

Although controversy continues over the causes of the Arabian Sea monsoon, agreement exists over its predictability. The southwest monsoon da Gama followed to India occurs when the land heats up faster than the ocean during summer. Because the ocean absorbs more heat than the land, it remains cooler. As the land temperature rises above India, the air above it expands, creating a low-pressure region. The cooler ocean retains high pressure, so the moist ocean blows toward the land.

After moving to the main trading harbor of Calicut, the account tells of a small Portuguese party being greeted by two Tunisian merchants. After getting over their astonishment at meeting two Castilian and Genoese speakers who understood Portuguese, a brief exchange followed. The text characterized the Tunisians' surprised exclamation as expressing, "What the devil! Who brought you here?" The meeting remained friendly, as could only occur with strangers discovering others speaking the same language in a remote location.

A MEETING WITH THE RULER OF CALICUT

Following a practice of contacting unfamiliar communities developed over decades, da Gama informed himself of local greeting rituals, which he observed upon meeting the *zamorin* (ruler) of Calicut. As in previous encounters of Portuguese captains with newly encountered peoples during the reign of Prince Henry, da Gama carried letters from the Crown proffering friendship with the Indian leader and offering to establish a trading relationship. One of the difficulties appeared in the language of the letters, prepared in Portuguese and Arabic. After realizing that the language of the court was Persian rather than Arabic, later visitors (following established Portuguese practice) undertook to learn Persian.[52]

On this voyage, however, negotiations proved unsatisfactory, ending with the hereditary monarch refusing da Gama's final offer of exchange: four sweatshirt-style hoods, six hats, four bundles of coral, a bundle of basins containing six pieces, a box of sugar, and four full barrels, two of olive oil and two of honey, in exchange for samples of spices to take back to Portugal. The zamorin mocked da Gama's offerings and seized the men accompanying Diogo Dias, who had been sent with a second offer, locking them up in the building that contained the merchandise the Portuguese intended to exchange for spices and precious stones.[53] To sidestep official channels, yet acquire the spices he sought, da Gama sent members of his crew into the town, where they succeeded in purchasing cloves, cinnamon, and gems from street vendors. He rescued his men by capturing six high-ranking local inhabitants, who had made their way to the Portuguese ships anchored in the harbor, and exchanging them for his own. With ships from the zamorin in pursuit, da Gama took off to a safe harbor at some distance from the Indian court.

13

A Dreadful Mistake

The Return from India

Nearly all the boats returning to East Africa followed a route unique to this part of the world. Ships waited until the monsoon winds reversed and began to blow from the northeast, allowing them to return between November and February. Such monsoonal reversals occur elsewhere, but nowhere else do they involve a large stretch of open ocean between ports. As safe as it seemed, this traditional route had its perils, but it had been plied for centuries.[1]

However, da Gama chose to take another route that proved far more lethal than he could have known. First, why did he avoid the traditional path? Three factors may have colored his decision. First, he may have been thinking of the series of boats the zamorin of Calicut sent after him that had militarily attacked his fleet.[2] Had he waited for the monsoon reversal, da Gama would have needed to delay his departure in unfriendly territory for over three months until the winds changed. They might not have succeeded in remaining hidden from a large-scale attack during this time, a potential risk that may have led da Gama to reject the apparent route as more dangerous.

Second is the possibility, though unlikely, that da Gama and his crew were ignorant of the unique seasonal reversal of winds and currents in the Arabian Sea and the pilot who guided them from Africa to Calicut had neglected to inform them. The hostile ruler of Calicut would have had no interest in furnishing this information. Once they were safely far away from the Indian coast, an Arab on board their returning ships mocked their ignorance, saying that they were lost along the beach and did not even know how to return to their land. This potential ignorance suggests that da Gama may have lacked information about the monsoon reversals.[3] Finally, it is also possible that a more positive ambition could have motivated the decision to cast off. Lacking firsthand information on the nature of the Arabian coast, da Gama wanted to carry out a complete survey of the area.

The Tunisian merchants who first greeted da Gama in India had willingly described their return route, since it required a long, slow journey over land, one impossible for the Portuguese with their heavy ships and cargoes. Da Gama needed to sail south past East Africa, then around the continent's southern tip, and finally along the Atlantic to Portugal.[4]

Lacking a pilot with knowledge of the conventional return pattern, da Gama combined the information from local sources and applied his knowledge of the patterns of ocean currents to his route. Since he had sailed northeast on the way to the coast of India, the usual circular movement of ocean currents would predict that initially heading northwest would be wise. Cruising along a semicircular arc, a ship would eventually reach Mogadishu on the coast of Africa, from which he had departed. But sluggish currents and weak, variable winds conspired to make this trip slower than da Gama had anticipated.

Before setting out on his return, da Gama needed to pause for several weeks to prepare the ships for the journey. Thus, he secured a temporary haven among a group of six tiny, uninhabited islands 232 nautical miles (430 kilometers) northwest of Calicut in an area today occupied by the Indian navy. The largest island, Angediva, comprises 0.6 square miles (1.5 square kilometers). It contained nothing of value to the local communities, who rarely visited from their homes along the coast, about a nautical mile (1.8 kilometers) away. In addition to the absence of potentially hostile residents, Angediva held an abundant source of fresh water. Da Gama had all three ships anchor alongside this island, where they began to undertake the cleaning and repairs necessary for the return trip to Africa—as well as scavenging a captured Arabian ship for parts.[5]

Before departing on their return to Africa, the crew demonstrated their resourcefulness by trading with the local community for the necessary supplies. They acquired plentiful firewood for cooking, salted local fish to supplement what they could catch at sea, and immense stores of fresh water, having learned a bitter lesson from the African coast. The sailors also managed to secure vegetables, including cucumbers and pumpkins. Despite their yellow color, the local pumpkins contained only one-fifth the vitamin C of an orange, a crucial shortcoming that the sailors could not have realized. Once fully provisioned (minus the unknown vitamin), the sailors repaired the sails and cleaned the hulls, showing their strategic planning and readiness for the journey ahead.[6]

This delay spelled problems, though. They set off to the northwest along the coast of India too early to catch the return winds, and departed at the worst possible time. The southwest monsoon winds had begun to weaken, and the winds slowed as if resting between two powerful sessions. Ignorant of this, da Gama set out on a journey of over two thousand miles at a sluggish pace.[7] Their only advantage lay in the freshwater supply in wells along the southern Arabian coast.[8] On many days, the winds remained still, increasing the anxiety of all on board. After eighty days at sea, da Gama signed a pact with the two other captains in his fleet, indicating a collective decision. The pact stated that if the wind did not pick up in fifteen days, they would attempt a return to India, a desperate measure that would almost certainly have ended in disaster, because the winds were about to make the seasonal shift to flow in the opposite direction.[9]

The stretch of ocean that nearly drove da Gama to abandon the journey was dominated for much of the year by winds and currents spinning like a demented top that was 167 nautical miles (309 kilometers) across. As it spun, it wobbled back and forth between an island due south of the Arabian Peninsula and Somalia on the east coast of Africa. As autumn progressed, the top began to turn slower and slower, until it finally stopped. This wind current is called the Great Whirl, a significant natural phenomenon in the Indian Ocean. However, the arrival of the northeast winds that the ships so desperately needed depended on the complete collapse of the Great Whirl. The unpredictable timing of its depletion seemingly conspired with another aspect of the northwest Indian Ocean to slow the fleet's progress.[10]

Winds and currents throughout the Arabian Sea reverse direction, usually during fall. However, the timing varies by region. Across the southern edge of the sea, winds and currents began to flow southwest in November, encouraging Arab and Indian voyagers to head to Africa. However, along the sea's western boundary, where da Gama was sailing, these same forces shifted much later in the year as the exhausted top temporarily blocked the arrival of monsoon winds from the northeast.

A second factor also delayed the change in direction of the currents that da Gama's fleet desperately needed. Beneath the surface of all the world's oceans flow long Rossby waves, planetary waves that move slowly westward. Unlike other waves, Rossby waves can disperse energy before or after the wave's arrival. This Rossby wave would slow the start of the southwestward flowing currents off the northern Somali coast until at least December, or occasionally January.[11] Thus, da Gama was almost ready to give up the journey when the northeast monsoon winds and currents finally appeared on December 27, 1498.

They had been just under two weeks away from losing everything.[12] Once blessed with favorable winds and currents, the fleet advanced more swiftly. On the second of January 1499, the crew saw land for the first time since leaving India. The sun had already set, so they anchored offshore overnight. The following day, a large urban center came into sight: Mogadishu, one of the largest cities on the East African coast and a major commercial center. As the winds forced the fleet to pass close to the harbor, da Gama took the unusual maneuver of firing broadsides to warn off any ships whose officers might have been considering an attack on the distinctively rigged Portuguese fleet.

By the time they reached Mogadishu, the fleet had lost at least one and possibly two crucial crew members: their onboard mapmaker and their trained astronomer, whose names went unmentioned in the account despite their importance. These deaths prevented the fleet from mapping the coast or determining their latitude, as a single individual often carried out both tasks. Their absence left the crew unsure of their precise position and, therefore, uncertain of the distance to the friendly port of Malindi. Without knowing the exact length of their journey, they navigated cautiously, sailing only during the day to avoid accidentally passing the friendliest port they had encountered along the Arabian Sea. Despite only sailing during the daytime, the fleet traveled to Malindi at a swift 5.7–6.6 knots. On January 7, 1499, only five days after passing Mogadishu, they finally reached Malindi harbor. By this time, they had been at sea for ninety-four days, the longest and deadliest portion of the journey.

Their Arab and Indian counterparts had enjoyed a far speedier trip. Instead of the three months that da Gama and his crew suffered through, local ships arrived on the East African coast from India in less than three or no more than six weeks. Da Gama's more circuitous route took eighty-nine days to sight land, just one day shorter than the entire transatlantic crossing, and turned out to be more deadly for the crew, for a reason that has rarely been noted: the small amounts of vitamin C in the pumpkins acquired in Angediva.

The yellow interior of the light green pumpkins native to India contained only twelve milligrams of vitamin C per hundred grams. By contrast, the average orange contains 53.8 milligrams. Even adding a small amount of the vitamin by eating cucumbers (4.5 milligrams/100 grams) would only have allowed the sailors to receive 30 percent of their daily requirement. Thus, although returning across the Arabian Sea took only one day more than crossing the Atlantic, thirty men died of scurvy in the same timeframe. Upon reaching friendly territory along the African coast, da Gama's first request to the local leader was for a supply of oranges. He immediately distributed the fruit to stricken crew members, but by then, the ailing men could not consume enough fruit to reverse the course of the disease.

This complicates our understanding of how knowledge of scurvy evolved. It is generally understood that a British doctor connected citrus to scurvy prevention only in the eighteenth century.[13] However, Vasco da Gama and those provisioning his fleet in 1497 clearly understood that eating citrus fruit was the way to prevent scurvy during that ninety-three-day transatlantic journey. After all, none of his crew suffered or died from the disease during that time; only when their supplies ran out did they start to develop scurvy, as they lacked the knowledge of local foods to choose fruit that would protect the crew's health.

The loss of these thirty men to scurvy was keenly felt. In addition to the deaths of the mapmaker and astronomer, many ordinary seamen also perished, leaving only seven or eight crew to operate each ship. After spending five days in Malindi, the now understaffed fleet searched for a secure location to reconstitute itself. Not wanting the population of Malindi to observe their diminished numbers, nor wishing for any of their technology to be captured and used, they managed to sail south for a single day. After twenty-four hours they reached Mtangata Reef, just offshore of the second most northern harbor in today's Tanzania. This port was not part of the Arabian Sea chain of trading posts and held only local fishing boats.[14]

On January 13, 1499, they began to take apart the *São Rafael* after lightning struck the ship during a sudden thunderstorm two days before they arrived in Malindi. The stays (long ropes underneath the sails) were damaged beyond repair, making the ship hard to maneuver. Even though Captain Nicolau Coelho often took the lead along the South African coast, the fleet commander and his brother Paulo captained the other two vessels, contributing to the decision to scrap the *São Rafael*. After salvaging the sails, supplies, navigational equipment, and other crucial parts of the ship, the sailors burned the remaining timbers so as not to leave behind any material that enemies of Portugal might reuse.[15]

The crews spent fifteen days off the Mtangata Reef, disassembling the *São Rafael*, cleaning the hulls of the remaining ships, and preparing for the next phase in their journey. During this time, they traded with the residents of Cape Mtangata, about one nautical mile away, exchanging shirts and armbands for chickens and hens. The remaining crewmen were divided between the other two vessels, fourteen on one and eighteen on the other, providing enough workforce to handle each boat for the remainder of the voyage. On January 27, the ships *São Gabriel* and *Berrio* left Mtangata for the long journey home.[16]

The account races through the remainder of the journey, as if wishing for it to end quickly. On January 18, they passed by Zanzibar, the giant island off the coast of Tanzania. On February 1, they arrived at Goa Island in Mozambique, where they offloaded the last pillar they had been entrusted with. Pouring rain the following day prevented them from soldering the cross to the top of it, so they left the pillar behind without the cross and departed without waiting for better weather.

A month-long silence followed, with subsequent remarks appearing only after they had sailed through the Mozambique Channel and the treacherous Agulhas to the halfway point around Africa's southern coast. Upon reaching Saint Blaize Bay, halfway between the Cape of Good Hope and Port Elizabeth, they paused to catch penguins, sea lions, and anchovies, which they salted to preserve for the rest of the journey. The only remaining major challenge rested with the passage around the Cape of Good Hope.[17]

On their first attempt to depart from Saint Blaize, they sailed briefly before a strong east wind forced them to retreat to the harbor. However, beginning on the next day, March 13, favorable winds returned for almost a week, letting them head west to the South Atlantic. Since they had only recently come from a warm climate, they shivered in the cold winds around the cape, which seemed far more biting than they had remembered. However, they succeeded in rounding the cape without incident.

Doubtless anxious to return to Portugal as soon as possible, they again used their early understanding of the Coriolis effect once in the South Atlantic. Instead of heading back across the ocean to Brazil (the way they had arrived), they followed the powerful winds and currents moving in the opposite direction from those on the western side of the Atlantic. The northward-heading Benguela Current was the perfect vehicle for a speedy northward return voyage. In only twenty-seven days they had left the swift Benguela behind, and from there the pilots knew enough about the remainder of the route used to return from the

Guinea coast to make an easy trip home despite landing in a region of light and variable winds.[18] Nine days later, they reached the Bijagos archipelago, where the travel account ends as the journey passed through areas well sailed for over fifty years. Pilots followed the well-traversed route west to the Cape Verde Islands, north to the Azores, and back to Lisbon.[19]

By April 16, 1499, as the fleet reached Cape Verde's port of Santiago, Paulo da Gama had become extremely ill. Vasco da Gama asked Coelho, whose ship had been burned off the coast of Tanzania, to take over his brother's vessel and return to Lisbon as soon as possible while da Gama remained behind. Vasco da Gama then took his brother to the warm, balmy weather of the Azores while Coelho took Paulo's ship, the *Berrio*, to Lisbon, where he made land on July 10, 1499. Coelho's arrival was greeted with great jubilation by King Manuel I, who lost no time in dispatching messages to the Crown of Spain and to Pope Alexander. These letters celebrated the return voyage in florid terms, signaling the king's intent to claim sovereignty over this route to the Indies and gloating over the lengthy list of rare spices the fleet had acquired in India.

Shortly after Vasco da Gama sent his ship back to Lisbon, where it arrived shortly after the *Berrio*, he chartered a smaller vessel to take his brother to the mild climate of the Azores. Shortly thereafter, Paulo passed away. Devastated by his beloved brother's death, da Gama spent over six weeks in the islands mourning his passing, until at last he summoned the emotional strength to return to what he knew would be a hero's welcome, all while his own heart was broken. It was a bittersweet homecoming for a man who had led one of the most significant maritime expeditions in the history of humankind. Even after leaving the Cape Verdes, instead of proceeding directly to the capital, da Gama spent nine days at a monastery on the outskirts of Lisbon, delaying his arrival until September 8, almost nine weeks after the first ship returned.[20] He had succeeded in an epoch-making journey, but it had meant losing his brother.

14

The Salty Tears of the Atlantic

O salty sea; how much of your salt is tears?

—Fernando Pessoa, "Mar Português," 1934

Countless families grieved over sailors who set out from Lisbon or Lagos and never returned. Likewise, those enslaved persons who became victims of Iberian trade in the Atlantic were mourned back home. Soon, that grief would be shared by millions, as the horrors of triangular trade spread from country to country, creating the colonial world that would define the centuries to come.

Perhaps this bitterness suits the Atlantic, whose surface contains more salt than any other ocean, because less rain falls to dilute the brine. Its captives' tears undoubtedly cast a broad net over the heartbreak of the Atlantic's history.[1] Thus, the story of the first maritime routes across the Atlantic and the connection between two of the world's oceans by sea cannot be told truthfully without a narrative synthesizing heroism, cowardice, compassion, and cruelty.

By continuously calculating geometrical angles, the first ship to go around the Bulge found a sea route no earlier sailors had managed to navigate. Through sheer determination and unwavering courage, subsequent sailors mastered the spinning North Atlantic gyre and southern Gulf Stream, an achievement they followed by deciphering the complex crisscrossing of the equatorial currents and recognizing the opposite directions of currents flowing north and south of the equator. Finally, they invented the first mathematically accurate technique to calculate latitude, thus forever shattering the barrier to sailing in the southern seas and inaugurating an area of global sea travel that would alter humanity's understanding of the planet as an oceanic world.

Among the many remarkable achievements of fifteenth-century Portuguese navigators, one has endured longer than the rest: their ingenious ability to define the significant sea lanes for Atlantic and southern Indian Ocean travel. Da Gama pioneered routes for crossing the Atlantic in 1497, a feat that continues to inspire awe and admiration. These routes were followed by later explorers from Francis Drake to James Cook, and they are the same seaways that modern container ships and tankers use today.[2] Despite his harsh and intolerant personality, Bartolomeu Dias identified the route for sailing around the eastward bend of South Africa and into the Indian Ocean. When modern tankers cannot use the Suez Canal, they travel the return route discovered by Dias. These men were, quite literally, trailblazers, identifying the critical routes that allow our modern global economy to exist.

Sailing, especially in uncharted waters, relied on more than instruments, maps, and written instructions. To grasp unfamiliar circumstances, sailors needed constant vigilance by all their senses. Journeys far overseas began in a world of winds and waves, safe harbors, tempest-driven seas, rocky coasts, and shifting sands. Fogs resting on the sea or giant banks of clouds, sometimes scurrying, sometimes lingering overhead, blocked all light, leaving sailors to depend on their visual, aural, and whole-body senses. The ability to read the sea, to see and correctly judge the heights of waves, to sense the direction of swells and speed of currents, to hear a shift in the direction and velocity of winds before they tore through the sails, to recognize subtle changes in sounds of the bow as it struck the waves, and to feel the rolls and swells of the blue-gray ocean beneath their feet allowed nautical pioneers to steer their boats safely across thousands of miles of uncharted seas.

By the end of the fifteenth century, knowledge obtained from these journeys would start undermining an ancient and widely held belief that the world consisted principally of land. Crossing the Atlantic proved the existence of a sea between the Atlantic and Indian Oceans, providing the foundation for understanding how little of this world is solid ground. The definitive blow to this land-centered view of our planet also came from a Portuguese navigator: Fernão de Magalhães (also known as Ferdinand Magellan), who proved that a waterway connected the Atlantic and Pacific Oceans and that a vast ocean linked this passageway to Asia, using the knowledge of the Coriolis effect established decades before to make his case. Indeed, our modern understanding of the ocean world began with Portuguese pilots and navigators, even if we do not and will never know most of their names.[3]

These unique and remarkable achievements cannot be separated from the horrors inflicted on Indigenous peoples of Africa by the Portuguese, their leaders, and the European powers that followed in their wake. Prince Henry's whip-wielding separation of families was an unforgivable cruelty. Two Indigenous communities, Denkira and Akan, successively enslaved thousands of their fellow residents along the Guinea Coast, only to endanger their lives by sending them to work in unsafe gold mines. Humans enslaved and tortured other humans in order to secure the economic incentives created by these breakthrough voyages. Cruelty and greed do not have a gender, religion, color, or nationality.

There were participants of all sorts in this unseemly exchange, from greedy merchants and traders to self-important captains and pilots who thought nothing of causing harm to people who had never done anything to hurt them. Later, successful and wealthy merchants left the organization and cultivation of goods and the capture of people to those local leaders willing to participate and exploit their community for sales in new markets.

SOFT POWER

As brutal as the slave trade was, the Portuguese Crown also practiced what we now call soft power on the African continent, co-opting rather than coercing. One strategy brought the children of essential rulers to Lisbon for education: "The king also has in his court many sons of lords from Africa who are instructed in our religion and laws."[4] This unique treatment of the offspring of powerful rulers provided both sides with familiarity during commercial or diplomatic exchanges.

Second, the king looked to create alliances through gifts to individuals in local positions of power:

> And he [the king] immediately sends them [local leaders] presents, with which he appeases them so that his agents can now traverse many lands on the continent of Africa and investigate each one. Formerly, he had traded only around the . . . coast, and they [his representatives] did not presume [that they could] enter many miles inland on their own.[5]

Finally, he used his subjects' appreciation for foreign languages to encourage Portuguese to learn the languages of Indigenous peoples and to deploy enslaved interpreters to initiate contact with potential trade partners and expand further searches within Africa: "The king also has blacks . . .

with different languages who know Portuguese. Using [these blacks] as two-way interpreters, [the King's agents] can walk around all of Africa."[6]

The combination of education for African elites, gift exchanges, and interpreters smoothed the way for continued commercial exchanges. These steps also allowed Portuguese travelers to venture safely into unfamiliar territory, searching for additional commercial goods in Africa's interior.

Critics of Portuguese overseas conduct have often claimed that these gifts and exchanges proved that the Portuguese were either ineffectual or subjected to local leaders. However, one foreign observer noted that such actions suited Portuguese economic interests exactly: "And [the King] appeases the major rulers with gifts, for it is impossible for him to rule over them. Even if he had dominion, it would be useless to him."[7] Thus, the Portuguese enlisted local leaders who desired income or power from acquiring the variety of goods transported from Portugal; consequently, these Indigenous collaborators took charge of the means of production and mobilized and organized the labor force. Portuguese kings continually rewarded the cooperating African leaders who could manage their groups in harvesting crops (Malagueta pepper), collecting elephant tusks, mining gold, or capturing people.

Despite the hands-off policy of soft power, Portuguese trade often disrupted Indigenous societies' traditional hierarchies. By organizing community output and trading with the Portuguese, interloping captains could help friendlier African societies disturb the rule of their historical overlords, replacing traditional leaders (as did the Budomel clan in Senegal) or becoming independent. Thus, encouraging the replacement of inconvenient local rulers with more market-friendly ones became an enduring practice of Western colonizers.

Allowing native communities to organize and manage the collection of goods and people guaranteed a level of internal independence. However, with that reconstituted social order came increased, oversized demands for their products, damaging and disrupting society and the environment as new intermediaries and leaders hastened the pace of labor to meet unprecedented demand.

Exploitative, yes, and brutal, but a different arrangement than under the later Spanish and English empires, which seized land and imposed novel forms of government and religion. The Portuguese welcomed those Indigenous communities that converted, but these visitors were more interested in trade than in spreading their Catholic faith, as the Spanish subsequently did in the Americas. Nor did Portuguese princes send hordes of their fellow citizens to seize the natives' lands, as did the English. However, that does not mean that the actions of Portuguese colonialists were morally better. Everywhere, the Empire of Exchange pushed the exploitation of native labor to new levels.

This empire was the forerunner of countless overseas colonial enterprises on islands and continents worldwide, originating in Sweden, Denmark, France, England, the Netherlands, Belgium, and Germany—uniformly inhumane in their treatment of other peoples. All of these enterprises and empires were made possible by the rapid pace of nautical innovations in the fifteenth century, ranging from sailing practices to the science of celestial navigation, methods still taught to ordinary sailors and trainees in all the world's significant navies.

The final Portuguese achievement uncovered a direct route across the Atlantic to the Americas. Without Portuguese nautical innovations, Europeans would only slowly have become acquainted with the landmasses to the west. Settlement would instead have begun by moving southward from Canada (easily accessible from Greenland) down the eastern coast of the United States, only gradually revealing the size of two continents. Portuguese navigational knowledge made the explosion of immigration to the New World in the fifty years after Columbus's arrival in the Caribbean possible. The histories of settlement in the Americas we know so well would have occurred very differently had the Portuguese captains, navigators, pilots, and crews not opened the Atlantic to Europe.

Notes

Acknowledgments: George Marcus, Patrick Manning, Kim Ricker, Antônio Tomás, Noah Zachary, the University of California Press, and Massachusetts Institute of Technology Fellowship in Science, Technology, and Society Department. Art assistance: Arielle Zepeda. Maps by the author, except maps 10, 11, and 13 by Matt Drwenski.

ABBREVIATIONS

Historical primary sources are cited throughout the notes using the following abbreviations. See the bibliography for full source information.

Africa Pilot	UK Hydrographic Office, *Africa Pilot*
Barros	Barros and de Couto, *Décadas da Ásia*, vol. 1
Brásio	Brásio, *Monumenta missionaria Africana*
Cadamosto	Cadamosto and Sintra, *Viagens de Luís de Cadamosto e de Pedro de Sintra*
CVF	Fernandes et al., *Códice Valentim Fernandes*
Esmeraldo	Pereira and Barradas de Carvalho, *Esmeraldo de situ orbis de Duarte Pacheco Pereira* (slightly different copies of this source appear in Évora and Lisbon, as noted in the citations)
PMA	Albuquerque et al., *Portugaliae monumenta Africana*
Velho	Velho et al., *Diário da viagem de Vasco da Gama*
Zurara	Zurara and Soares, *Crónica dos feitos notáveis . . .*

INTRODUCTION

1. Predictable seasonal reversals occurred along northern Indian Ocean basins, not short-term reversals in smaller sections of these currents. For example, the Bay of Bengal has a northward current along the Indian shelf from January through October and a southwestward current the rest of the year. See Matthias Tomczak and J. Stuart Godfrey, *Regional Oceanography* (Oxford: Pergamon, 1994); Rhodes Whitmore Fairbridge, *The Encyclopedia of Oceanography* (New York: Reinhold, 1966); S.R. Shetye et al., "Hydrography and Circulation in the Western Bay of Bengal During the Northeast Monsoon," *Journal of Geophysical Research: Oceans* 101, no. C6 (1996): 14011–25. For tenth-century Muslim attempts, likely southward down Morocco's coast, see Évariste Lévi-Provençal, *Histoire de l'Espagne Musulmane (710–912)* (Cairo: Institut français d'archéologie orientale du Caire, Études historiques, 1945), vol. 1, iii, 342n.

2. Divisio describing an argument by Fabianus, in Lucius Annaeus Seneca, *The Suasoriae of Seneca the Elder; Introductory Essay, Text, Translation and Explanatory Notes; Being the "Liber Suasoriarum" of the Work Entitled L. Annaei Senecae Oratorum et Rhetorum, Sententiae, Divisiones, Colores*, ed. William A. Edward (London: University Press, 1928), *Suasorio* 1, sec. 10. See also Duane W. Roller, *Through the Pillars of Herakles: Greco-Roman Exploration of the Atlantic* (New York: Routledge, 2006).

3. Al-Masudi, in his enigmatic work *Muruj adh-Dhahab*, describes the Atlantic: "No ship sails therein, nor is any habitable land there, nor any reasonable creature dwelling therein. Neither its extent nor end is known." Quoted in Lévi-Provençal, *Histoire de l'Espagne Musulmane*, vol. 1, 157, 218, 224, 393, 441; Michel Bochaca, *Voyager par mer au Moyen Âge: Navigations dans le Ponant au XV*[e] *Siècle* (Paris: Editions Gisserot, 2023).

4. Seneca, *The Suasoriae of Seneca the Elder, Suasorio* 1, sec. 16; Al-Masudi, quoted in Lévi-Provençal, *Histoire de l'Espagne Musulmane.*

5. Martha W. Buckley and John Marshall, "Observations, Inferences, and Mechanisms of the Atlantic Meridional Overturning Circulation: A Review," *Reviews of Geophysics* 54, no. 1 (2016): 5–63.

6. Luís Adão da Fonseca, *O essencial sobre Bartolomeu Dias* (Lisbon: Imprensa Nacional-Casa da Moeda, 1987). For da Gama's voyage, see Sanjay Subrahmanyan, *The Career and Legend of Vasco da Gama* (Cambridge: Cambridge University Press, 1997); Genevieve Bouchon, *Vasco de Gama* (Lisbon: Fayard, 1997).

7. Polynesians navigated primarily in the Southern Hemisphere, venturing north of the equator to Hawaii. For details, see chapter 1.

1. NATURE INTERVENES

1. "[Hubo] três fases: uma primeira com a duração aproximada de minuto e meio, pouco violenta, intervalo de um minuto, um movimento mais intenso durante cerca de dois minutos e meio, provocando já danos consideráveis; outro intervalo de um minuto, seguindo-se a terceira e última fase por três minutos, a mais violenta de todas. Total, cerca de nove minutos." Joachim José Moreira de Mendonça, *História universal dos terremotos, que tem havido no mundo, de que ha noticia, desde a sua creaçaõ até o seculo presente: Com huma narraçam*

individual do terremoto do primeiro de novembro de 1755, e noticia verdadeira dos seus effeitos em Lisboa, todo Portugal, Algarves, e mais partes da Europa, Africa, e América, aonde se estendeu: e huma dissertaçaõ phisica sobre as causas geraes dos terremotos, seus effeitos, differenças, e prognosticos; e as particulares do ultimo (Lisbon: Antonio Vicente da Silva, 1758), 11, 113–14, 117, 120, 285–93, 160. See also António Pereira de Figueiredos, *Commentario Latino e Portuguez Sobre o Terremoto e o Incêndio de Lisboa*, 8 vols. (Lisbon: n.p., 1756); Anon., *Breve compendio de las innumerables lamentables ruìnas, y lastimosos estragos, que à la violencia, y conjuracion de todos quatro Elementos, experimentò la Gran Ciudad, y Corte de Lisboa, el dia primero de noviembre de este año de 1755* (Barcelona, Spain: Imprenta de Teresa Piferrer viuda, 1755); Miguel Tibério Pedegaches Ivo, *Nova e Fiel Relação que Experimentou Lisboa, e Todo Portugal no. 1 de Novembro de 1755*, and Manuel Portal, *Historia Da Ruina Da Cidade Cauzada Pello Espantozo Terremoto e Incendio Que Reduzio a Pó e Cinza a Melhor e Maior Parte Desta Infeliz Cidade* (1756, publisher names and locations unknown). For a contrary opinion on the height of the tsunami, see P.-L. Blanc, "Earthquakes and Tsunami in November 1755 in Morocco: A Different Reading of Contemporaneous Documentary Sources," *Natural Hazards and Earth Systems Sciences* 9 (2009): 725–38. Even with his lower figures, the navigational material would have been lost.

2. Joseph Fowke, ed., *A Genuine Letter to Mr. J. Fowke, from His Brother, Near Lisbon, Dated November 1755, in Which Is Given, a Very Minute and Striking Description of the Late Earthquake* (London: n.p., 1755); Maria Leonor Machado de Sousa et al., *O Terramoto de 1755, Testemunhos britânicos / The Lisbon Earthquake of 1755, British Accounts* (Lisbon: British Historical Society of Portugal, 1990).

3. Well into the nineteenth century, Atlantic Ocean pilots preferred to use handwritten logs and maps to ease editing. This resulted in the loss of working records, leaving only lavishly painted court documents. For working nineteenth-century pilots, see John Harris [1792–1802] and Richard Hussey Moubray [d. 1803] in Cunninghame Graham [Angus, Sir], Naval History Manuscripts Collection 1746–1967, University of Irvine Special Collections, MS.M.001. The Portuguese began printing occasional nautical guides only toward the end of the sixteenth century.

4. *Esmeraldo de situ orbis* (ca. 1507), first published in 1892. Duarte Pacheco Pereira and Joaquim Barradas de Carvalho, *Esmeraldo de situ orbis de Pacheco Pereira / édition critique et commentée [par] Joaquim Barradas de Carvalho* (Lisbon: Fundação Calouste Gulbenkian, Serviço de Educação, 1992 [1892]) (hereafter cited as *Esmeraldo*, either the Lisbon or the Évora copy).

5. Comissão Executiva das Comemorações do Quinto Centenário da Morte do Infante D. Henrique and Comissão Executiva das Comemorações do Quinto Centenário da Morte do Infante D. Henrique, *Monumenta Henricina* (Lisbon: n.p., 1960). Valentim Fernandes copied three texts by hand, only one of which was a navigational guide, and Alvise Cadamosto summarized only the most memorable moments of his twin voyages and appended a slightly shaky second-hand account of Pedro de Sintra's expedition. Luís de Cadamosto and Pedro de Sintra, *Viagens de Luís de Cadamosto e de Pedro de Sintra* (Lisbon: Academia

Portuguesa de História, 1988). The earliest misplaced nautical guides were copies of the *Esmeraldo situ orbis* by Pereira. The two scientific texts were by Abraham ben Samuel Zacuto and Joseph Vecinho: *O Regimento da declinação do sol . . . tirada . . . do Zacuto pelo honrado Gaspar Nicolas* (Geneva, Switzerland: [Société Sadag], 1917), and *Almanach perpetuum celestium motuum (Radix 1473): Tabulae astronomicae* (Leiria, Portugal: J. B. Obernetter, 1496).

6. For example, João de Barros et al., *Da Asia de João de Barros e de Diogo de Couto* (Lisbon: Na Regia officina typografica, 1777); Fernão Lopes da Castanheda, *Ho livro primeiro dos dez da historia do descobrimento & conquista da India pelos portugueses* (Coimbra, Portugal: por Ioao da Barreyra, 1554).

7. Eric Ash, *Power, Knowledge, and Expertise in Elizabethan England* (Baltimore: Johns Hopkins University Press, 2004), 89; Leyla Perrone-Moisé, *Vinte luas: viagem de Paulmier de Gonneville ao Brasil, 1503–1505* (São Paulo, Brazil: Companhia das Letras, 1992); Alida Metcalf, *Mapping an Atlantic World, Circa 1500* (Baltimore: Johns Hopkins University Press, 2020).

8. The idea of an uninhabitable torrid zone appeared in classic Greek and Roman works, including Strabo and Ovid, and was repeated (and occasionally contested) during the Middle Ages. Ovid, *Ovid's Metamorphoses*, ed. Madeleine Forey, trans. Arthur Golding (Baltimore: Johns Hopkins University Press, 2002), bk. 1. A medieval map appeared in William Harris, *Macrobius Commentary on the Dream of Scipio* (New York: Columbia University Press, 1990).

2. AROUND THE BULGE

1. *Africa Pilot*, vol. 1, chap. 9, sec. 236; chap. 5, sec. 265. Some earlier nautical guides mistakenly characterized the sardines as waves.

2. "The continental shelf of the whole region is the most extensive of any eastern boundary current." Javier Arístegui et al., "Oceanography and Fisheries of the Canary Current/Iberian Region of the Eastern North Atlantic," in *The Sea*, vol. 14, ed. Allan R. Robinson and Kenneth H. Brink, 877–931 (Cambridge, MA: Harvard University Press, 2004), 878. For an overview, see Robert E. Todd et al., "Global Perspectives on Observing Ocean Boundary Current Systems," *Frontiers in Marine Science: Ocean Observations* 6 (August 2019).

3. "Cabo de Bojador he huũ cabo ae no mar e tambem se chama Bojador Caba d'Área," *CVF*, 39.

4. "E porq~ ata este cabo foy a nauegaçã dos castellanos e portugueses e deste cabo auãte nunca se achou alguẽ q~ nauegasse pellas grãdes corretes," *CVF*, 39.

5. Victoria Murden McClure, personal communication, 2001.

6. On May 28, 2021, a ship caught in the Canary Current washed up on the Caribbean island of Tobago with forty-three fatalities. Renata Brito and Felipe Dana, "Adrift in the Atlantic, a Boat of Death and Lost Dreams," Associated Press, April 11, 2023, https://apnews.com/article/adrift-investigation-migrants-mauritania-tobago-debe63712438 61fb287a2e3a2df6727b. For similar stories in Brazil and Barbados, see Renata Brito, "9 Bodies Found Adrift in a Boat Off Brazil Were Likely Migrants from Mauritania and Mali, Police Say," Associated Press, April 16, 2024; Giles Tremlett, "After Four Months at Sea Ghost Ship with Eleven Petrified Corpses Washes Up in Barbados," *The Guardian*, May 28, 2006.

7. Brito and Dana note the number of similar fatalities in 2021 (the most recent year for which statistics are available) as between 1,109 and 4,000. "Adrift in the Atlantic."

8. Angela Schottenhammer, "The 'China Seas' in World History: A General Outline of the Role of Chinese and East Asian Maritime Space from Its Origins to c. 1800," *Journal of Marine and Island Cultures* 1, no. 2 (2012): 63–86; Ben Finney, *Pacific Navigation and Voyaging* (Honolulu: Polynesian Society, 1976); Ben Finney, *Voyage of Rediscovery: A Cultural Odyssey through Polynesia* (Berkeley: University of California Press, 1994); Christophe Picard, *La mer et les musulmans d'occident au moyen age Viie-Xiiie siècles* (Paris: Presses Universitaires de France, 1997), 20–21.

9. For Norse attempts to cross Bojador, see *The Saga of the Volsungs, The Saga of Ragnar Lodbrok, Together with the Lay of Kraka*, trans. Margaret Schlauch (New York: American-Scandinavian Foundation/W.W. Norton, 1930). For the king of Mali, see al-'Umari in *Corpus of Early Arabic Sources for West African History*, trans. J.F.P. Hopkins and Nehemia Levtzion (Princeton, NJ: Markus Wiener, 2011), 268–69.

10. "Sailing rapidly, we passed by a fiery region filled with vapors, from which great torrents of fire flowed down to the sea. The land could not be approached because of the heat. We sailed away from there quickly, being struck with fear. Then, having sailed for four days, we sighted a land full of flames at night. And in the midst of it there was a fire higher than the rest which seemed to touch the stars. By day we discerned it to be a mountain of great height." Introduction in Jacques Ramin, *Le Périple d'Hannon / The Periplus of Hanno* (Oxford: British Archaeological Reports, 1976). See also Gilda Risica et al., "Reconstruction of the Subaerial Holocene Volcanic Activity Through Paleomagnetic and 14C Dating Methods: El Hierro (Canary Islands)," *Journal of Volcanology and Geothermal Research* 425 (2022). None of the other volcanic islands—Cape Verdes, the Cameroon Hot Line–São Tomé–Principe Chain—erupted at any time during the Phoenician navigation of the Mediterranean. Franck Eitel Kemgang Ghomsi et al., "Main Structural Lineaments of the Southern Cameroon Volcanic Line Derived from Aeromagnetic Data," *Journal of African Earth Sciences* 186 (2022); Dan N. Barfod and J. Godfrey Fitton, "Pleistocene Volcanism on São Tomé, Gulf of Guinea, West Africa," *Quaternary Geochronology* 21 (2014): 77–89; J.D.A. Piper and A. Richardson, "The Palaeomagnetism of the Gulf of Guinea Volcanic Province, West Africa," *Geophysical Journal International* 29, no. 2 (1972): 147–71. Early skeptics of Hanno's claims included Pliny the Elder. Joyce Irene Whalley, *Pliny the Elder, Historia naturalis* (London: Sidwick & Jackson, 1982), 5.1.

11. Seneca the Elder (54 BCE–ca. 39 CE). Lucius Annaeus Seneca, *The Suasoriae of Seneca the Elder, Suasorio* 1, sec. 10.

12. Al-Mas'udi warns, "Beyond me [the pillars of Hercules] is no route nor way for those who would enter the yon sea from the Sea of the Greeks [i.e., the Mediterranean]." Muruj Al-Mas'udi, *L'Abrégé des merveilles*, trans. Carra de Vaux (Paris: Forgotten Books, 1898), 365. Al-Idrisi described the sea off the English and French coasts as follows: "Ces pays étant baignés du côté du couchant par la mer Ténébreuse, il vient continuellement de ce côté des brumes, des

pluies, et le ciel est toujours couvert, particuliérement sur le littoral. Les eaux de cette mer sont épaisses et de couleur sombre les vagues s'y élévent d'une maniére effrayante; sa profondeur est considerable. L'obscurité y régne continuellement." Muḥammad Al-Idrisi, *Geographie, d'Edrisi*, 2 vols., trans. Amedée Jaubert (Paris: Imprimerie royale, 1836–1840), vol. 2, 355–56.

13. The exact number of years is disputed, although it was over a decade. "Mandou lá bem xiv vezes ates que soube parte da dicta terra." *PMA*, 1, 23 (October 22, 1443). See José Ramo Coelho, ed., *Alguns documentos do Archivo Nacional da Torre do Tombo ácerca das navegações e conquistas portuguezas publicados por ordem do governo de sua majestade fidelissima ao celebrar-se a commemoração quadricentenaria do descobrimento da America* (Lisbon, Imprensa Nacional: 1892), 8. Three other sources report twelve years of failed journeys: Zurara, 52–53, "doze anos"; *CVF*, 39; *Esmeraldo* (Évora), fol. 40v.

14. Fernando Pessoa, "Mar Português," poema X do Canto Sexto "Possessio Maris," in *Messagem* (orig. pub. 1934).

15. Such ships likely carried no more than seven people. Zurara, 81–82. No complete firsthand information on the dimensions of Portuguese vessels appears before 1580. Francisco Contente Domingues, *Os navios do mar oceano: teoria e empiria na arquitectura naval portuguesa dos séculos XVI e XVII* (Lisbon: Centro de História da Universidade de Lisboa, 2004), 24.

16. Zurara, 52–55.

17. For pilots unable to carry out the simple geometric calculations involved in zigzagging as they sailed into the wind, a cheat sheet called a *toleta de marteloio* was developed, likely during the late fourteenth century or possibly later. This valuable tool provided a quick reference for the necessary calculations, enhancing the efficiency of navigation. See Franco Masiero, "La raxon de marteloio," *Studi Veneziani* 8 (1984): 393–412; H. Michea, "De l'Utilisation de Martelloios ou Roses des Vents au Moyen Age," *Navigation*, no. 111 (1989): 101–17. For a skeptical assessment, see Ramon Jose Pujades i Bataller, *Les cartes portolanes: la representació medieval d'una mar solcada* (Barcelona, Spain: Institut cartogràfic de Catalunya, 2007), 463–64. *CVF* refers to a zig-zag pattern but does not mention the need to consider the current's speed.

18. Eanes likely utilized the "current triangle," a remarkable tool that corrected the direction of travel. Its three legs, representing the ordered course (C) and speed (S), the current's estimated set (S) (direction) and drift (D) (velocity), and the resultant track (TR) and speed of advance (SOA), made solving the problem for the third leg of the triangle possible. This ingenious tool was a significant leap in navigation. For a more detailed analysis of current sailing off Bojador, see Patricia Seed, "Navigating the Mid-Atlantic; or, What Gil Eanes Achieved," in *The Atlantic in Global History, 1500–2000*, ed. Jorge Cañizares-Esguerra and Erik R. Seeman, 77–89 (Upper Saddle River, NJ: Pearson Prentice Hall, 2007).

19. A different technique used in competitive sailing today demands the ability to see landmarks in the distance, a task that becomes significantly more challenging when viewing the shifting sands of the Sahara.

20. *Fagonia cretica* (called Rosas de Santa Maria in Portuguese) produces purple flowers only between December and June, making these months the

most likely time for this voyage and underlining the importance of this natural indicator in navigation.

3. FROM MISTAKEN EXPECTATIONS TO CONQUESTS

1. Zurara, 55.

2. See map 1.

3. Zurara, 58.

4. *Africa Pilot*, vol. 1, chap. 6, sec. 32.

5. At the entrance to the "Rio," Baldaia found an estimated five thousand "Lobos marinhos," most likely monk seals, whom he described as easy to kill and whose fur he could carry on the boat. "Fez matar aquelles que pode de culas pelles fez carregar su navio." Zurara, 57, 59–60. Zurara records the king's age as six. *Africa Pilot*, vol. 1, chap. 5, sec. 259.

6. Antonio Cordeyro, *História insulana das ilhas a Portugal sujeitas no oceano ocidental*, 2 vols. (Lisbon: Tipografia do Panorama, 1866), bk. 4, chap. 2, 135; Gago Coutinho, "Significação Nautica do Centenario da Descoberta dos Açores," *Seara nova*, nos. 1587–1598/1599 (1978/79): 359–61. For a history of the many controversies about the discovery of the Azores, see Artur Teodoro de Matos et al., eds., *História dos Açores: do descobrimento ao século XX* (Angra do Heroísmo, Portugal: Instituto Açoriano de Cultura, 2008), 49–59.

7. Gilles Reverdin et al., "North Atlantic Ocean Surface Currents," *Journal of Geophysical Research: Oceans* 108, no. C1 (2003): 3002; Lothar Stramma and Thomas J. Müller, "Some Observations of the Azores Current and the North Equatorial Current," *Journal of Geophysical Research* 94, no. C3 (1989): 3181–86. In 1419, Portuguese sailors were blown off course, only to encounter Canary Islanders fishing along the shores of islands just 295 nautical miles (546 km) north of the Canaries. Six years later, the Portuguese began to settle on the island, which they named for the tall, straight tree trunks useful for masts—*madeira* (wood) in Portuguese. However, despite being only three hundred nautical miles north of the Canary Islands, Madeira lies too far south of the Azores Current to facilitate a return to Portugal. Camille Parrain, "Sailing Routes and Stopovers: Spatial Disparities Across the Atlantic," *Journal of Coastal Research* 2011, no. 10061 (2011): 140–49.

8. Zurara, 68.

9. The second capture might have occurred in the Gulf of Cintra, a large (twenty-five-mile) half-moon-shaped bay sixty-five miles south of Dakhla on the coast of Río de Oro province.

10. Captures: Zurara, 62–64, 67–68.

11. Zurara, 68. The status of a knight (cavalryman) was the first step in raising a man's status through awards for bravery in battles.

12. "He certo que aquy há gente é que trautam com camellos. . . . E o trêfego daquestes deue sser contra o mar." Zurara, 62.

13. The Portuguese would later label this Berber language Zenaga, which is nearly extinct today.

14. Arabic translators worked in the Portuguese court through at least 1528. João de Sousa, *Documentos Arábicos para a história portuguesa* (Lisbon:

Academia Real das Ciências, 1790). According to Corriente, they "were still an overwhelming majority in certain areas, craft and trade, and underworld activities. . . . They consequently exerted considerable influence on the areas of language most closely connected with those activities." Federico Corriente, *Dictionary of Arabic and Allied Loanwords: Spanish, Portuguese, Catalan, Galician, and Kindred Dialects* (Leiden, The Netherlands: Brill, 2008), introduction.

15. "O primeiro que fez presa em esta conquista." In characteristic Iberian fashion, Zurara justifies this action using Augustine's *City of God*, which he claims describes such actions as accepted Roman practice, thus eliding the question of its more direct origins. Zurara, 65. For other examples of such elisions, see my essay "The Requirement," in Patricia Seed, *Ceremonies of Possession in Europe's Conquest of the New World, 1492–1640* (Cambridge: Cambridge University Press, 1995), 69–99.

16. For a different interpretation, see Paul H. Hair, "The Early Sources on Guinea," *History in Africa* 21 (1994): 87–126. This work provides a contrasting view on the early sources of Guinea, which is relevant to our discussion of the historical narratives surrounding slavery and colonialism.

17. For Portuguese titles distinguishing "crusade" from "conquest," see Rui de Pina, *Chronica d'El-Rei D. Affonso V*, 3 vols. (Lisbon: Escriptorio, 1901–2): "conquista" (vol. 1, 144; vol. 3, 16, 17, 133) vs. "cruzada" (vol. 2, 133, 140–41, 143); Rui de Pina, *Chronica d'El-Rei D. Duarte* (Lisbon: Escriptorio, 1907): "conquista" (59, 67, 68, 72, 81, 87, 119) vs. "cruzada" (6, 62, 63, 93, 99, 105, 122); and Damião de Góis and Gabriel Pereira, *Chronica d'El-Rei D. Manuel*, 8 vols. (Lisbon: Escriptorio, 1909): "conquista" (vol. 1, 26, 37, 65; vol. 2, 26, 40, 48; vol. 4, 9, 109, 125–26; vol. 5, 34, 77) vs. "cruzada" (vol. 3, 53; vol. 8, 21).

18. The convent was Convento de las Trinitarias Descalzas. Cervantes wrote about his experiences as a slave in *El trato de Argel, La Galatea*, and *Los trabajos de Persiles et Sigismunda*. See Maria Antonia Garces, *Cervantes in Algiers: A Captive's Tale* (Nashville, TN: Vanderbilt University Press, 2002); David Santillana and Istituto per l'Oriente, *La comunità musulmana e il suo capo, fonti del diritto e loro ermeneutica, la legge nello spazio e nel tempo, le persone, la famiglia, i diritti reali* (Rome: Anonima Romana Editoriale, 1926), 143–44.

19. Garces, *Cervantes in Algiers*, 19–40.

20. An evolution of this phrase decades later suggests that the Portuguese slavers were rescuing prisoners from death by poison arrows during battles and selling them back to their friends. While such rescues may have occurred occasionally, they were dwarfed (even in this author's text) by the 40,000 ducats the king received annually from sales of enslaved people in Spain and Italy. Frederick Kunstmann, *Hieronymus Münzer's Bericht über die Entdeckung der Guinea* (Munich: Verlag der K. Akadamie), 356–57.

21. "Filhar gente per qual quer maneira que milhor podesse. . . . Presa mais rica por que o senhor Iffante possa cobrar alguũ começo de paga sobre tãta despesa"; "[V]iemos a esta terra por leuar carrega de tam fraca mercadorya acertamas agora ẽ nossa dicta de leuar os primeiros cautivos ãte a presença do nosso principle." Zurara, 62–68. Charles Verlinden seizes upon this statement as the first evidence of a profit motive for slavery: "Favorisés d'une meilleure chance, nous en trouvions quatorze ou quinze, dont nous ferons une prise de plus

grand profit'. Profit! Le mot est lâché et il est d'importance capitale. D'ailleurs le commentaire de Zurara montre bien que l'homo economicus était à l'affût dans l'âme de ces compagnons des premiers découvreurs." Charles Verlinden, "Les débuts de la traite portugaise en Afrique (1433–1448)," in *Miscellanea mediaevalia in memoriam Jan Frederik Niermeyer* (Groningen, The Netherlands: J.B. Wolters, 1967), 365–78. However, the profit motive had existed for centuries in the mutual enslaving of Muslims and Christians in the Maghreb.

22. Traditional servitude: Zurara, 77.

23. Muhammad Akram Khan, *Islamic Economics and Finance: A Glossary* (London: Routledge, 2003), 60.

24. Traditional servitude: Zurara, 77. On Arabic spoken in Portugal then, see Maria Filomena Lopes de Barros, "A minoria muçulmana do reino português e os contactos diplomáticos com o dār al-islām," in *Comunicación y política en la baja Edad Media*, ed. Neéstor Vigil Montes (Évora, Portugal: Cidehus, 2019).

25. Zurara, 77–80. Gonsalves reportedly told the prince he would seek information about Asia and a mythical African Christian named Prester John.

26. "Passasse aallẽ do porto da gallee o mais lõge que pudesse." Zurara, 66. Arrival at Cabo Blanco: Zurara, 69. The *Bula Etsis suscepti* (see note 37) suggests that Henry had taken these captives. For an overview of the essential primary sources in English, see Hair, "The Early Sources on Guinea," 87–126.

27. *Africa Pilot*, vol. 1, chap. 5, sec. 259; chap. 6, sec. 8.

28. South of Cabo Blanco, upwelling is limited to winter and spring due to the northward migration of the Azores high during summer, which is responsible for driving equatorward winds. Pierre-Marie Pulain, "Coastal Ocean Circulation Experiment off Senegal," https://www.onr.navy.mil/reports/FY11/popoulai2.pdf.

29. Patricia Seed, "The Impact of Rising Sea Levels on Banc d'Arguin Wildlife Refuge," in *Esri Map Book 2010* (Redlands, CA: Esri Press), 45.

30. Contemporary Arabs referred to them by a variety of names. Al Qalqachandi (1418 CE) called them "the Iẓnagen / Sanhaja (Zenaga)." Aḥmad ibn 'Alī Qalqashandī et al., *Two Studies on the Geography and Administration of Egypt According to al-Qalqašandī (d. 1418)*, ed. Fuat Sezgin Mazen Amawi et al. (Frankfurt: Institute for the History of Arabic-Islamic Science at the Johann Wolfgang Goethe University, 1992).

31. *CVF*, 42.

32. All of this coast belongs to modern Mauritania, which became the last nation in the world to abolish slavery in 1981. In spite of its official termination, however, slavery continues, as numerous overseas-based organizations of escaped enslaved people continue to fight for its elimination in practice, not merely in law. Catherine Coquery-Vidrovitch, *Les routes de l'esclavage: histoire des traites africaines*, vol. 20 (Paris: Albin Michel, 2018). See also SOS Esclaves, a Paris-based NGO, https://sos-esclaves.com/; Kwame Anthony Appiah and Henry Louis Gates Jr., *Encyclopedia of Africa* (Oxford: Oxford University Press, 2010), 549; Chouki El Hamel, "Race, Slavery and Islam in Maghribi Mediterranean Thought: The Question of the Haratin in Morocco," *The Journal of North African Studies* 7, no. 3 (2002): 29–52; Martin A. Klein and

Suzanne Miers, *Slavery and Colonial Rule in Africa* (London: Routledge, 2013), 58–59, 79–86.

33. Carta de quitação a Diogo Afonso Malheiro. *PMA*, 31–32.

34. Zurara, 122.

35. Zurara, 81–82. Letters of both 1443 and 1448 refer to newly created maps of this coast. "E mandou dela fazer carta de marear"; "Y mandou dello fazer carta de marear." *PMA*, 23 and 25.

36. Pope John XXII (elected in 1410) had to flee a council meeting for the safety of the territories of the Duke of Austria. The pope in Avignon survived only with the protection of the kings of Castile, Aragon, and Navarre. Pope Martin V, elected in 1417, had to flee Rome and hide in Florence for several years because of conflict with the Duke of Milan. The classic study of the popes in this period is Joseph Esmond Riddle, *The History of the Papacy to the Period of the Reformation* (London: Richard Bentley, 1854), vol. 2, 350–73 and 501–2. The payment reference appears in Zurara, 73.

37. "Nobilis viri Henrici, ducis Visensis . . . temporalibus regere gubernateque valet . . . quod terras, possessiones et alia mobilia et immobilia bona quaecumque in regnis ac dominiis Regis Portugaliae, et quibuslibet alüs locis consistentiaque praeatae militiae per quosvis Christi fideles donari vel alias per am justis modis acquiri." Shortly after Tristão's ships returned with seventeen captives, on January 9, 1442, Eugene IV (1431–47) issued the *Bula Etsis suscepti*, a papal bull granting Prince Henry sole authority. The complete Latin text appears in Pius Onyemechi Adiele, *The Popes, the Catholic Church and the Transatlantic Enslavement of Black Africans 1418–1839* (Hildesheim, Germany: Georg Olms, 2017), 500–501. The envoy was Fernão Lopes. Zurara, 73.

38. "Licet tune personaliter non intersit, milites, . . . necnon ali Christi fideles plerique . . . bella contra Sarracenos ac inimicos praefatos mover, . . . exercitus hujusmodi bellis, ac conflictibus interfuerint . . . de quibus corde contriti, et ore confessi extiterint, auctoritate Apostolica, . . . concedimus remissionem peccatorum." Eugene IV, *Illius qui se pro divini* (papal bull granting forgiveness), January 14, 1442. (Emphasis added.) The full text appears in Adiele, *The Popes, the Catholic Church*, 501–2. The payment reference appears in Zurara, 73.

39. "Nulli ergo omnino hominum liceat hanc paginam nostrae concessionis infringere, vel ei ausu temerario contraire." Eugene IV [1421–47], *Illius qui se pro divini*, in Adiele, *The Popes, the Catholic Church*, 501–2. Zurara (1) summarized Eugene's two decrees in Portuguese as if they were a single order, (2) added the language of Nicholas V's *Dum diversas* regarding the legitimacy of capturing infidels, and (3) dropped the harsh language of the consequences of opposition to include only the final words of *Dum diversas:* "incur the wrath of God and the apostles Peter and Paul." Zurara, 74. The regent of Portugal almost simultaneously authorized Henry to receive his customary 20 percent share. Zurara, 74–75; Nicholas V (1397–1455) *Dum diversas* [papal bull], June 18, 1452.

40. Zurara, 85, 87. Lançarote's ships had approximately 2.3 times the capacity of Tristão's. See note 42 below for total slave numbers. J. Semedo de Matos, "A caravela de Nuño Tristão: A caravela de descobrir," *Revisita de Armada*, no. 218 (1997), 21. The caravel was first mentioned in 1441. Its innovations included two lateen sails and a cable for mooring lintels to masts, allow-

ing vessels to sail closer to the wind. The increased capacity did not occur until Lançarote's expedition.

41. Zurara, 81–82.

42. They arrived on the feast of Corpus Christi, which occurs sixty days after Easter. Zurara, 87. In 1444, Easter occurred on April 14. Hence, Corpus Christi occurred on June 13 or June 14 if they celebrated the feast on a Thursday. Women bearing turtles: Zurara, 95. Sleeping child kidnapped: Zurara, 88–89. In three cases, Zurara is unsure of the number by only one. Adding all the captives and taking the three uncertain ones yields a total of 240. In one case, however, he declares that forty-eight people were taken but accounts for only forty. If forty-eight is correct, the maximum would have been 245–48. In any case, 93–95 percent of those captured survived. The number of survivors is based on the size of Prince Henry's fifth plus the two given to the church. Zurara, 106, 109.

43. Islamic Maliki jurisprudence is divided on this issue, with some authorities adopting the Shia convention, which states that the fifth goes to the imam. By contrast, other Sunni schools assigned the fifth to a secular leader such as the emir. In the Christian Iberian tradition, the fifth (*quinto del botín*) always went to the lay leader. Alfonso IX and Gregorio López, *Siete partidas*, Partida 5, Title 26, Law 6; Halil ibn Ishāk, *Il lMuntasar "Sommario del diritto malechita,"* trans. Ignacio Guidi and David Santillana (Milan, Italy: Ulrico Hoepli Ditore Libraio della Real Casa, 1919), vol. 2, 395–96, 402–6; Location and timing of division: Zurara, 105–6.

44. Zurara, 105.

45. "Onde couijnha de necessydade de sse apartarem os filhos dos padres e as molheres dos maridos e os huús irmaãos dos outros. / a amigos nem a parentes nom se guardaua nhúa ley. soomente cada huù caya onde a ssorte leuaua." Zurara, 108. One of the captains had died during the journey, leaving only five to divide the remaining 80 percent.

46. "E uos outros que vos trabalhaaes desta partilha esguardaae com piedade sobre tanta miseria e veede como se apertã huûs cô os outros / que apenas os podees deslegar." Zurara, 108. Jerome Branche has noted the description of the brutality in his "Inscribing Contact: Zurara, the Africans, and the Discourse of Colonialism," *Palara* 2 (2001): 6–19.

47. "Os filhos que vyà os padres na outra / alleuâtauanse rijamête e hyanse pera elles / as madres apertauà os outros filhos nos braços e lâçauanse cô elles de bruços / recebendo feridas cò pouca piedade de suas carnes por lhe nò seerê tirados." Zurara, 108.

48. "O câpo era todo cheo de gête assy do lugar como das aldeas e comarcas darredor . . . soomète por veer aquella nouidade / E cònestas cousas que vyâ / huús chorado outros departindo / fazyam tamanho aluoroço que poinham è toruaçõ os gouernadores daquella partilha." Zurara, 108–9.

49. "O Iffante era ally en cima de huù poderoso cauallo / acópanhado de suas gentes repartindo suas mercees come homê que de sua parte querya fazer pequeno thesour / Ca de Rvj. almas que acontecerom no seu quinto / muy breue fez delles sua partilha. ca toda a ssua principal riqueza staua em sua vootade / consijrádo cò grande prazer a saluaçò daquellas almas que ante erã perdidas." Zurara, 108–9.

50. Zurara, 109.

51. "Fazeremos seruiço a deos e ao Infante nosso senhor." Zurara, 87. After the first biggest attack and killing, Zurara wrote: "Enfim nosso senhor deos que a todo bë da remuneração quis que pollo trabalho que tijnhã tomado por seu seruiço aquele dia cobrassë vytorya de seus jmigos. E galhardamente e paga de seus trabalhos e despesas." Zurara, 89.

52. Zurara, 89.

53. Zurara, 87.

54. Two previous papal declarations had referred to a Christian militia, but only in passing: *Illius qui se pro divini* (1442) mentions tactics of "the militia mentioned above" among other justifiable actions; *Bula Etsis suscepti* (1442) refers to Henry's role as a leader of "the army of Christ" and to his subjects as members of the same Christian militia: "Praefatae militiae" in *Illius qui se pro divini* and "Magistratus militia Jesu Christi" and "Magistro ac fratribus ejusdem militia" in *Bula Etsis suscepti*. Adiele, *The Popes, the Catholic Church*, 501–3.

55. "Christi nominis inimicorum rabies . . . , Christianaeque Religioni valet subjugari . . . illius inimicos potenti student brachio expugnare." *Dum diversas*, in Adiele, 503.

56. "Sarracenos, Paganos, infideles, et Christi inimicos . . . subjugandi, illorumque personas in perpetuam servitutem redigendi." Adiele, 503.

57. This document also cut Prince Henry out of sole ownership of all the possessions taken from "Muslims, pagans, infidels, and enemies of Christ." It shifted these profits to the kings of Portugal and their successors in perpetuity. "Sarracenos, Paganos, infideles, et Christi inimicos . . . tibi et successoribus tuis Regibus Portugalliae, perprtuo applicandi, et appropriandi, ac in tuos, et eorundem successorum usus, et utilitates convertendi plenam, et liberam, auctoritate Apostolica, tenore praesentium concedimus facultatem, eandemque Regiam Majestatem."

58. "Carta de quitação," May 14, 1444. *PMA*, 1 and 31. Six weeks after *Dum diversas*, Pope Calixto III affirmed the right of Crown-appointed representatives to "trade with the unbelievers and the Saracens to make purchases and sales of whatever goods and supplies it would befit them . . . provided they would not carry over (iron) tools, wood, ropes, ships, or any other kinds of armaments." *PMA*, 1 and 74.

59. Nicholas VI's 1452 decree had only mentioned conversion in passing. "Qua ad hujusmodi salutiferum opus dicta videlicet defensionem, augumentationem que Religionis cooperari conspicimus." *Dum diversas,* in Adiele, 503. Because *Romanus pontifex* was signed in January 1455, Henry and Afonso would have needed to send the request at the end of 1444.

60. *PMA*, 23; Zurara, 294.

61. Anon., Cantino chart, Biblioteca Estense, Modena, Italy. By da Gama's return in 1499, the map of Africa was completed, with material from the Americas added in August 1502. For a detailed account of its acquisition see Alida C. Metcalf, *The Mapping of the Atlantic World* (Baltimore: Johns Hopkins University Press, 2019).

62. "Ad ipsius Infantis pervenisset no quam nunquam, vel saltem a memoria hominum, non consuevisset per hujusmodi Occeanum mare versus meridionales et orientales plagas navigari, illudque nobis Occiduis adeoforet incogni-

tum." Nicholas V, *Romanus pontifex* [papal bull], January 8, 1455, "eisque incognitum sacratissimum Christi nomen predicare" (to preach the most sacred name of Christ [which was] unknown to them).

63. Nicolas V, *Romanus pontifex*.

64. Nicolas V, *Romanus pontifex;* Zurara, 108–9.

65. Zurara, 71.

66. Pope Eugene IV's *Bula Etsis suscepti* and *Illius qui se pro divini* explicitly owe their existence to "noble Henry Duke of Viseu's request." The *Dum diversas* (1452) only reserved perpetual slavery "to the Kings of Portugal and your successors." *Dum diversas*'s complete phrasing is "illorumque personas in perpetuam servitutem redigendi, regnaquoque, Ducatus, Comitatus, Principatus, aliaque Dominia, possessiones, et bona hujusmodi, tibi et successoribus tuis Regibus Portugalliae, perpetuo"; "nobilis viri Henrici Ducis Visensis, petitio." *Illius qui se pro divini*, 501, 503. *Dum diversas* addresses "nobilis Vir, Henricus, Infans Portugalliae, Charissimi in Cristo Filii Nostri Alphonsi, Portugalliae et Algarbii Regnorum Regis illustris," and *Bula Etsis suscepti* "Nobilis viri Henrici, ducis Visensis." *Inter caetera* of Pope Callixtus III, 1456. "Sicut pro parte Alphonsi Regis et henrici Infantis predictorum." Both the January 1455 *Romanus pontifex* and the 1456 *Inter caetera* open with greetings to "the noble Henry, Infante of Portugal . . . [and] illustrious king Alfonso of the kingdoms [of] Portugal and the Algarve."

67. See previously cited texts of *Illius qui se pro divini, Bula Etsis suscepti*, and *Dum diversas*.

68. For uses of "conquest," see note 17. For the fourteenth-century origins of the word "crusade," see the etymologies and first definitions of *croiserie* and *crusade* in the *Oxford English Dictionary*. The individuals that Adahu exchanged for his freedom were likely non-Muslim "Azenagues," a traditional slave community that exists to this day. In his flawed *Origem da Lingoa Portugueza*, Duarte Nunez de Liao declares that the Portuguese meaning of *conquista* is unique (Lisbon: Pedro Crasbeeck, 1606), 124.

69. On its disappearance, see Barros, *Décadas* 1, bks. 1–5. Critics who have noted plagiarized sections of the chronicle include Joaquim de Carvalho, *Sobre a Erudição de Gomes Eanes de Zurara (notas em torno de alguns plágios deste cronista)* (Coimbra, Portugal: Universidade de Coimbra, 1949); Luís Filipe Barreto, *Gomes Eanes de Zurara e o problema da "Crónica de Guiné"* (Lisbon: STVDIA, 1989); Alfredo Pinheiro Marques, *Vida e Obra do Infante D. Pedro* (Lisbon: Gradiva, 1996), 86–87. Critics of the laudatory tone include Duarte Leite, *Acerca da "Crónica dos Feitos de Guinee"* (Lisbon: Bertrand, 1941); Vitorino de Magalhães Godinho, *Documentos sôbre a Expansão Portuguesa* (Lisbon: Gleba, 1945); Bailey W. Diffie and George D. Winius, *Foundations of the Portuguese Empire, 1415–1580* (Minneapolis: University of Minnesota Press, 1977), 48n; Peter Russell, *Prince Henry 'The Navigator': A Life* (New Haven, CT: Yale University Press, 2000), 5.

4. HOW TRADING REPLACED CONQUEST

1. Cadamosto, 122.

2. The earliest slave traders are listed by *Esmeraldo* (Évora), fol. 104–5.

3. The word for warehouse, *feitoria*, is often misconstrued as meaning "factory." But *feitoria* denotes "a clear space in the woods where people spend the night and store supplies and clothes," formerly a "large commercial establishment." *Dicionário Priberam da Língua Portuguesa*, s.v. "Feitoria" (accessed February 14, 2023).

4. Arguin was the largest island in the bay south of the White Cape (Cabo Blanco) and well sheltered by the cape. In June and July, currents returning northward appeared just west of the bay. U.S. Hydrographic Office, *Atlas of Pilot Charts North Atlantic Ocean (Including the Gulf of Mexico)*, sec. 1.

5. Cadamosto (18) provides more realistic annual slave trade figures from Arguin than Damião Peres, who tries to minimize the volume of the slave trade. Damião Peres, *História dos descobrimentos portugueses*, 4th edition (Porto, Portugal: Vertente, 1992), 182.

6. The caravans carrying gold and slaves came to Arguin later. Magalhães Godinho, *Os descobrimentos e a economia mundial* (Lisbon: Editorial Presença, 1981), vol. 1, 79–86, 145–55.

7. The dominant green trees are *Acacia raddiana talh*. The traffic at Arguin is described extensively in Robert Ricard, "Le commerce de Berbérie et l'organisation économique de l'empire portugais aux XVe et XVIe siècles," *Annales de l'Institut d'Etudes Orientales de la Faculté des Lettres d'Alger*, vol. 2 (1936), 266–90. A shorter summary appears in A. Teixeira da Mota, *Some Aspects of Portuguese Colonisation and Sea in West Africa in the 15th and 16th Centuries* (Bloomington: African Studies Program, Indiana University, 1978), 4. According to Ricard, "cloth hambels, hafks, djellabas, horses, and North African wheat, all these used at the Arguin factory to acquire gold and slaves, which may have contributed to the conquest of Safin where the hambel industry held an important place." *Esmeraldo* (Évora) describes poor-quality red and blue cloth, cheap cloaks made in the Alentejo, and similar items of little value. *Esmeraldo* (Évora), fol. 105. The gold and the enslaved people arrived via caravans and also in the salt trade. E. Ann McDougall, "Salts of the Western Sahara: Myths, Mysteries, and Historical Significance," *The International Journal of African History* 23, no. 2 (1990): 231–57; Godinho, *Documentos*, 145–51; Alexandre Magno de Castilho, *Descripção e roteiro da costa occidental de Africa*, vol. 1, 123.

8. According to Godinho, *Documentos*, vol. 2, 147–48, João II built a failed warehouse in Ouadane or Wādān. Two forts were constructed to connect with the Guinea gold trade: Axim and Mina. The Portuguese abandoned Axim when travel between the gold fields proved onerous.

9. Ekkehard Mittelstaedt, "The Upwelling Area off Northwest Africa—A Description of Phenomena Related to Coastal Upwelling," *Progress in Oceanography* 12, no. 3 (1983): 307–31; Lothar Stramma and Friedrich Schott, "The Mean Flow Field of the Tropical Atlantic Ocean," *Deep Sea Research Part II: Topical Studies in Oceanography* 46, no. 1–2 (1999): 279–303; Josep Pelegrí and Aissa Benazzouz, "Coastal Upwelling off North-West Africa," in *Oceanographic and Biological Features in the Canary Current Large Marine Ecosystem*, ed. Luis Valdés and Itahisa Déniz-González, 93–103 (Paris: IOC Technical Series, 2015); T. Klenz et al., "Seasonal Variability of the Mauritania Current

and Hydrography at 18°N," *Journal of Geophysical Research: Oceans* 123, no. 11 (2018): 8122–37.

10. Ekkehard Mittelstaedt, "The Ocean Boundary Along the Northwest African Coast: Circulation and Oceanographic Properties at the Sea Surface," *Progress in Oceanography* 26, no. 4 (1991): 307–55.

11. When Cadamosto arrived at Budomel's territory, he observed that the Portuguese had already established trade in the region. Cadamosto, 36–37.

12. "Os primeiros negros mais chegados a nós são os Jalofus, que começão do Rio de Sanaga, da banda do Sul delle; este Rio 'os divide dos Alarves, que ficão da outra banda do Rio, da banda do Norte." André Alvares de Almada [active 1594] and António Brásio, *Tratado Breve Dos Rios de Guiné Do Cabo Verde Dês Do Rio Sanagá Até Os Baixos de Santa Ana de Todas As Naçoes de Negros Que Há Na Dita Costa e de Seus Costumes, Armas, Trajos, Juramentos, Guerras. Ano 1594* (Lisbon: Editorial L.I.A.M., 1654), chap. 1, 5.

13. McDougall, "Salts of the Western Sahara," 232. The Wolof Empire enslaved people long before the Portuguese arrived. Susan McIntosh, "Tools for Understanding Transformation and Continuity in Senegambian Society, 1500–1900," in *West Africa during the Atlantic Slave Trade: Archaeological Perspectives*, ed. Christopher de Corse (London: Bloomsbury Academic, 2001), 14–37, esp. 17–18; Almada and Brásio, *Tratado Breve Dos Rios de Guiné*, 8–10.

14. Robin Law, *The Horse in West African History* (Oxford: Oxford University Press, 1980). Law suggests they were vulnerable to the tsetse fly, while James Webb points out the variety of possible parasitic infections. James L. A. Webb Jr., "The Horse and Slave Trade between the Western Sahara and Senegambia," *The Journal of African History* 34, no. 2 (1993): 221–46; Ivana Elbl, "The Horse in Fifteenth-Century Senegambia," *The International Journal of African Historical Studies* 24, no. 1 (1991): 85–110; Almada and Brásio, *Tratado Breve Dos Rios de Guiné*, 8–10.

15. Almada and Brásio, *Tratado Breve Dos Rios de Guiné*, 4.

16. Zygmunt Catewicz and Ryszard Siwecki, "On Currents in the Coastal Zone of African Shelf off Saint Louis," *Oceanologia* 21 (1985): 59–75.

17. The most common evergreen at the cape was *Acacia nilotica* (or *Vachellia nilotica*), five to twenty meters high with a dense spheric crown and branches usually dark to black.

18. The existence of a Green Cape was first reported by Dinis Dias, an early explorer. However, the exact date of his arrival remains a subject of debate among historians, adding an element of mystery to the early accounts of exploration. Zurara, 128.

19. Zurara, 128–29.

20. Zurara, 282–84.

21. The motto was French: "Talent de bien faire." A previous expedition under Diniz Dias may have encountered the cape, but the surviving accounts are hazy. Barros, *Décadas* 1, bk. 1, chap. 16.

22. Silt around the base of mangroves provides homes for filter feeders such as coral, oysters, shrimps, and mussels. Elizabeth McLeod et al., "A Blueprint for Blue Carbon: Toward an Improved Understanding of the Role of Vegetated

Coastal Habitats in Sequestering CO2," *Frontiers in Ecology and the Environment* 9, 10 (2011): 552–60; Christian Nelleman et al., *Blue Carbon: The Role of Healthy Oceans in Binding Carbon: A Rapid Response Assessment* (Nairobi: GRID-Arendal United Nations Environment Program, 2009); Marc Simard et al., "Mangrove Canopy Height Globally Related to Precipitation, Temperature and Cyclone Frequency," *Nature Geoscience* 12 (2019): 40–45; Cheryl L. Doughty et al., "Mangrove Range Expansion Rapidly Increases Coastal Wetland Carbon Storage," *Estuaries and Coasts* 39 (2016): 385–96; Temilola Fatoyinbo and Marc Simard, "Height and Biomass of Mangroves in Africa from ICESat/GLAS and SRTM," *International Journal of Remote Sensing* 34, no. 2 (2013).

Since the seventeenth century, French writers have called these twenty-one rivers the "South Rivers" because they begin at Saloum, south of the Sahara. Salif Diop, *La côte Ouest-Africain du Saloum (Sénégal) à la Mellacoré (Guinée)* (Paris: L'Editions de L'Orstom, 1990), 24, 277. Paul Pélissier, *Les paysans du Sénégal. Les civilisations agraires du Cayor à la Casamance* (Saint-Yrieix-la-Perche: Imprimerie Fabrègue Hutchinson, 1966); U.S. Office of the Geographer, Bureau of Intelligence and Research, *Mauritania—Senegal Boundary,* International Boundary Study, 1967, 2.

23. From north to south, the significant rivers are Siné-Saloum, Diomboss, Bandiala, Gambia, Casamance, Sénégambia, Cacheu, Mansoa, Geba, Grande de Buba, Tombali, Cacine, Romponi or Mandiafara, Nunez, Mapatchez, Pongo, Konkouré, Somba, Soumbaya, Forecariah, and Mellacorée. The exact distance is 1,850 miles. Diop, *La côte Ouest-Africain du Saloum (Sénégal) à la Mellacoré (Guinée)*, 277.

24. The Serer generated the hydrogen cyanide by mixing the leaves of the unique local *Acacia macrostachya* Rchb. ex DC, *Gardenia triacantha* DC., *Asparagus africanus* Lam., and *Clerodendrum capitatum* steeped for ten days. See Joseph Kerharo and Jacques G. Adam, "Les plantes medicinales, toxiques et magiques des Niominka et des Socé des lies du Saloum (Senegal)," in "Plantes médicinales africaines," *Acta Tropica* 25, no. 5 (1964): 279–34. Present-day incoming tides travel between one and two knots. *Africa Pilot*, vol. 1, chap. 7, sec. 27. Water levels are highest during the rainy season (July–November), although tidal currents rarely exceed 0.6 knots.

25. The surviving cabin boys were André Dias and Álvaro da Costa, "ambos escudeiros do Infante, e naturais da cidade de Évora, dezanove [outros] morreram." Zurara, 224. While the river remains identified only by distance and not by a name, the people with the deadly arrows are named. The Nuño Tristão expedition "foram sua viagem até chegarem a além do Cabo Verde 60 léguas, onde acharam um rio." Zurara, 228. However, "De prima inventiona Guinee" states: "Et iecit in christianos sagittas de veneno et vulnerati sunt aliqui christiani / qui statim mortui sunt de veneno . . . et vltra procederunt ad vnam terram malorum hominum qui Serreos populi vocantur Serreos Et inuenerunt ex illis multos in littore maris cuns suis arcubus et sagittis venenosis." Only the Serer were known to possess instantly lethal poison. "De prima inventiona Guinee." *CVF*, 285–86.

26. Hydrogen cyanide poisoning leads to hemodynamic instability and cerebral edema. The above listed effects constitute a few of the symptoms of both. It is unclear if the ingredients of his unnamed domestic herbal concoction con-

tained any of the antidotes, namely a fermentation byproduct with nitric acid (amyl nitrate), a salt (sodium nitrate), and a sulfuric byproduct (sodium thiosulfate). If the concoction contained any of these three elements, Fernandes would have survived.

27. Zurara, 229.

28. Two nongovernmental organizations continue to fight slavery in Mauritania. Boubacar Ould Messaoud, a formerly enslaved person, founded SOS Esclaves in 1995, and Biram Ould Dah Ould Abeid created the Initiative for the Resurgence of the Abolitionist Movement in 2008.

29. "Ainda que as cousas seguintes nom foram trautadas com tanto trabalho e fortelleza como as passadas, cá despois deste anno avante, sempre si os feitos daquellas partes trautarom mais per trautos e aveença de marcadorya, que per fortelleza nem trabalho das armas." Zurara, 121. "[P]er uoler tratar con el paexe bona paxe e concordia o tor de lor beniuolentia." Cadamosto expressed similar sentiments: "[P]aexe de gambra [Gambia] da nuj tanto desiderata . . . che se per caso I Negri del paese vegnesseno con le lor almadie o sia barche ad asaltarli che subito se redssese al nauilio senza uoler con lor contender. E questo perche nuj eremo li conduti per uoler tratar cone l paexe bona paxe e concordia o tor de lor beniuolentia la qual se conueghia aquistarcon Inzegno e non per arme ne per forza." Cadamosto, 54. However, Prince Henry continued to advocate the conquest method, even though it had proven deadly. See *Dum diversas.*

30. Antoniotto Usodimare, "Carta de Antoniotto Usodimare," in *Viagens dos descobrimentos*, ed. José M. Garcia (Lisbon: Editorial Presença, 1983), 143.

31. "[A]uea habuto qualche noticia del nostro praticar con quelli negri de senega I quali non podea esser saluo catiuj hominj in uoler nostra amistade perche lor teniano per fermo che nuj altri christianj manzauemo carne humana e che comprauemo negri saluo per manzarli e che per questo non uoleano nostra amistade per algun modo del mondo." Cadamosto, 57–58. Diogo Gomes also adopted this new orientation, writing after the disaster with "Depois disto no seu conselho o Senhor Infante dizia que Iara os futuros não brigassem com aquela gente naquelas regiões, mas que tratassem alianças, e tratassem de comercio e com êles asséntassem pazes, porque a sua intenção era fazê-los christãos. E mandou caravelas preparadas para a paz e guerra." Diogo Gomes's translation in Vitorino Magalhaes Godinho, *Documentos sôbre a expansão portuguesa*, 3 vols. (Lisbon: Editorial "Gleba," 1943), vol. 1, 195. "O Infante en commendava muito aos Capitães, que não rompeffem guerra com os moradores da terra que defcubriffem, fenão mui forçados, e ifto depois de lhes fazer fuas amoeftações, e requerimentos da fé, paz, e amizade." Barros, *Décadas* 1, bk. 1, chap. 14.

32. Cadamosto, 54–55.

33. European scholars have characterized this shift slightly differently. Two prominent Portuguese writers differed more than others. Luiz Thomaz described "a change in the mentality of Prince Henry" from crusading to commercial interests. At the same time, Vitorino Magalhães Godinho countered that "a phase of regulated activity replaced a phase of unregulated activity devoted to the pursuit of instant gains." Luiz Thomaz, "Portugal," in Godinho, *Documentos sôbre a expansão portuguesas*, vol. 1, 151. Belgian historian Charles Verlinden and English historian Paul Hair seconded Godinho's opinion: "Contemporary sources

. . . mentioned economic gain freely." Paul H. Hair, "The Early Sources on Guinea," *History in Africa* 21 (1994): 87–126, esp. 9; Charles Verlinden, "Les débuts de la traite portugaise en Afrique (1433–1448)," *Miscellanea mediaevalia in memoriam Jan Frederik Niermeyer* (Groningen, The Netherlands: J.B. Wolters, 1967), 367. However, while participants acknowledged their commercial interests, they failed to mention the self-preservation tactics that would sacrifice enslaved people's lives instead of their own.

34. "[A] ley de Xpõ *tiinha que lhe provesse* seer em ajuda da guerra dos Mouros dAfrica, naqual el Rey dom Affonso, que entom regnava em Portugal, e elle em seu nome, com os outros seus vassallos e naturaaes, continuadamente trabalhavom" (emphasis added).

35. Patricia Seed, "Elefantordenens Portugiske Forbindelse," trans. Ole Magnus Mølbak, *Siden Saxo: Magasin for dansk historie* 23, no. 2 (2006): 14–23. The actual Danish name of the nobleman is impossible to figure out from its fifteenth-century Portuguese spelling. Teixeira da Mota identifies the most likely location of Vallarte's disappearance as a point somewhere slightly north of the Saloum. Avelino Teixeira da Mota, *Some Aspects of Portuguese Colonization and Sea Trade in West Africa in the 15th and 16th Centuries* (Bloomington: African Studies Program, Indiana University, 1978), 309. The area was Mandingo speaking, judging by the translator's ease of communication with the locals. "[L]evando pera ello dous naturaaes daquella terra por turgymaães." Zurara, chap. 64. João de Barros's account differs slightly in its details: "The blacks, as they promised him a price, said that they would soon bring him an elephant, where he saw it, and, having returned within three days, they came to call Balarte—saying to bring what they had promised. Balarte, having entered the ship's barge only with the sailors who were rowing him, reached land, and, on taking from the gourd of palm wine that a negro gave to a sailor, he leaned so far on the edge of the boat that the sailor fell overboard. And in their haste to pick up the sailor, they neglected the boat so that the waves came with it ashore because the sea was a little blistered. The blacks, seeing that ours could not be rescued from the ship, came upon them, from whom no more escaped than one who knew how to swim, who gave the reason for this case, and who, coming swimming, had looked behind him and seen that Balarte was in the stern of the boat, battling like a struggling man." Barros, *Décadas* 1, bk. 1, chap. 15.

36. Late that spring, the fleet's captain, Pedro Ernest Çarrabodes, died en route. *PMA*, 33, 38–39.

37. On February 5, 1451, Abrão de Paredes was allowed to trade in Guinea (*PMA*, 33); on February 27, 1453, Nuno Antunes de Gooes's two caravels were licensed to trade in Guinea beyond Rio São João (*PMA*, 38); late that spring, Captain Pedro Ernest Çarrabodes died while traveling to Guinea on this fleet (*PMA*, 39). Prince Henry supplied food for four caravels in 1454 (*PMA*, 79).

38. Antoniotto Usodimare, letter, December 12, 1455. Usodimare, "Carta de Antoniotto Usodimare," *Viagens dos descobrimentos*, 139–44. Further examples appear in *CVF*, 297. Texeira da Mota identifies other possible voyages between 1448 and 1455 but locates all destinations safely north of the Serer's poison arrows in the Sine-Saloum. Avelino Teixeira da Mota, "A Descoberta da Guiné," *Boletim cultural da Guiné Portuguesa*, no. 1 (1945): 318–21.

On Texeira da Mota's long residence in Guinea, see Paul H. Hair, "The Teixeira da Mota Archive and the Guinea Texts Project," *History in Africa* 10 (1984): 387–94.

39. Only strong tides would drive Diogo Gomes back to the Gambia River two years later (ca. 1457). He reached Elephant Island quickly (the only island in the middle of the Gambia, eighty-six nautical miles upriver). He continued to Kantora, an additional hundred nautical miles upriver, beyond which the ships could no longer travel. *CVF*, 178. While Kantora's position on trade routes has been well described, a second aspect of this location has not been: He found an area with a slave class at Kantora, and, as of 2021, slavery still existed in this region. At Koina, five land miles upriver from Kantora, there is a slave class called "Ganbanaax." Lamin Fatty, "Gambia: Judge in Koina Caste Scuffle Case Threatens to Revoke Accused Persons' Bail Following Their Absence from Court," *Foroyaa* (Serrekunda, Gambia), March 22, 2021.

40. Cadamosto, 62–63.

41. Cadamosto, 63.

42. The garment was a surcoat. Cadamosto, 64–65. Mandé was widely spoken in the area.

43. "E mandamoli a dir como Ieremo lo venuti per comandamento del nostro signor re de portagalo christam per far con luj bona amistade e per Intender da luj sel auea bixogno dele cosse di nostri paexi che ogni anno el nostro Signor Re Ie ne (de) mandarin." Cadamosto, 64. Malyn Newitt notes the shift but attributes it to a shift in interest, rather than a consequence of deadly encounters. Malyn Newitt, *Navigations: The Portuguese Discoveries and the Renaissance* (London: Reaktion Books, 2023), 120.

44. Glands near the civet's anus contain an oil used for centuries in perfumes.

45. The interpreters (all slaves) were ordered to have close contact with the natives. "Insieme determinissemo de mandar do dele nostre barche con li nostri trucimanj In terra a sauer cosse nuore del paese e el nome de questa fiumara e sauer chi era signor di questa parte." Cadamosto, 71. For more on the slave interpreters, see chapter 5.

46. Cadamosto, 72. By 1908, only the fifteen main islands out of eighty-eight had been surveyed. U.S. Hydrographic Office, *Sailing Directions for the West Coasts of Spain, Portugal, and Northwest*, 1908, 278–83.

47. Cadamosto, 72–73.

48. "[Q]uesto cauo e vn puocho piu alto chal tereno dela costa e mostraua la fronte de esso esser rossa e per questo li metemo nome el cauo rosso." Cadamosto, 72.

49. Today's Cabo Roxo marks the northern bank of the Cacheu River, six miles south of the cliffs and the only protected anchorage in the area. UK Hydrographic Office, *The African Pilot, or Sailing Directions for the Western Coast of Africa, vol. 1* (London: Hydrographic Office, 1856), 268; Diop, *La côte Ouest-Africain du Saloum (Sénégal) à la Mellacoré (Guinée)*, 277.

50. "[E] a questo fiume a nome il Rio di barbacinj e cosse e notada sula carta del nauigar fata per mj." Cadamosto, 52. Barabacins are between the Seres and Nomimans. *CVF*, 199.

51. *Africa Pilot*, vol. 1, chap. 7, secs. 131–38.

52. The location of the "Saint Anne" River is impossible to ascertain with any certainty. Because of the twenty miles of breakers along the coast (the Falulo, Barela, and Jufung Breakers and the Coimbra Bank), it could have been the Soobijack tidal creek of the British, now Essukujak (French), Bujejete (English), or possibly but less likely any one of several rivers running through Varela National Park. UK Hydrographic Office, *African Pilot, or Sailing Directions for the Western Coast of Africa* (1856), vol. 1, 5–60; Cadamosto, 72. Discouraging access to the Cacheu were the Cacheu and Garamas Spits, Cacheu Bank, and five breakers: the Cacheops, Cacheu, Parcel, Mata, and Domingo. *Africa Pilot*, vol. 1, 59–60, 65–75.

53. Cadamosto, 73; *Africa Pilot*, vol. 1, chap. 7, sec. 220.

54. Cadamosto, 72. By 1908, only the fifteen main islands, out of a total of eighty-eight islands, had been surveyed. U.S. Hydrographic Office, *Sailing Directions for the West Coasts* (1908), 268. The Rio Grande de Buba was once the most prominent river along the coast. Its eighteen-mile-wide mouth completely silted up by the 1990s. The other Rio Grande (Geba) was fifteen miles wide at its mouth. *Africa Pilot*, vol. 1, chap. 7, sec. 220.

55. "La mazor dele due almadie se acosto ala mia carauella e qui ne comenza a guardar per grandissima merauelia vedendone esser homini bianchi etiam guardauano la forma del nostro nauilio con lalbaro e con lahtena incrosada." Cadamosto, 72–73. "Desideroso de Intender questa generation li fixi parlar *ali mei trucimanj* di quali mai algun di lor pote Intender cossa che li disseseno ne meno *quelli dele altre carauelle*" (emphasis added). Cadamosto, 73.

56. "In questo lungo atrouassemo vna contrarietate che non se atroua In algun luongo doue nauiga christianj zoe chel se fa in quel paexe marea de acqua de montante e de zosente como fa a uenesia o in tuto el ponente." He asserted that high tides rushed in over four hours but receded over eight hours "la crese four e calla 8." Modern sources record a minor difference, with flood tides on the Rio Geba taking five hours and ebb tides seven, but Cadamosto remained infuriated by the unexpected speed and timing. Cadamosto, 73; *Africa Pilot*, vol. 1, chap. 7, sec. 170.

57. In the sand-filled bay, tides can reach 1.5 knots with frequent eddies. *Africa Pilot*, vol. 1, chap. 6, secs. 76–77. For the physics of tidal eddies, see C. Li et al., "Residual Eddies in a Tidal Channel," *Estuaries and Coasts* 29, no. 1 (2006): 147–58.

58. Patricia Seed, *The Oxford Map Companion: One Hundred Sources in World History* (Oxford: Oxford University Press, 2014), 70–71; Pero Falchetta, *Fra Mauro's World Map: With a Commentary and Translations of the Inscriptions* (Turnhout, Belgium: Brepols, 2006), vol. 5. Luís de Albuquerque maintains that the map proved disappointing to the Portuguese authorities, already out of date given advances made during its creation. Luís de Albuquerque, *Ciência e Experiencia nos Descobrimentos Portugueses* (Lisbon: Instituto de Cultura e Língua Portugues, 1983), 63–64.

59. Neville Chittick, "Kilwa and the Arab Settlement of the East African Coast," *The Journal of African History* 4, no. 2 (1963): 179–90. Archeological evidence showed that Arab coastal settlements began around 1000 CE but did not expand into the interior or venture below central Madagascar or Sofala.

5. LANGUAGE

1. There are plenty of records in other European languages for later centuries. But for the fifteenth, only occasional mentions remain in surviving Portuguese documents.

2. João de Sousa, *Documentos Arábicos para a história portuguesa* (Lisbon: Academia Real das Ciências, 1790), shows the extensive use of Arabic in Portugal. See Federico Corriente, *Dictionary of Arabic and Allied Loanwords: Spanish, Portuguese, Catalan, Galician, and Kindred Dialects* (Leiden, The Netherlands: Brill, 2008).

3. Zurara, 91. On September 2, 1448, King Afonso V stated that one thousand prisoners had been taken since rounding Bojador. *PMA*, 25. Subtracting the totals from the voyages Zurara detailed (232 plus approximately another 17 and 12) leaves at least an additional seven hundred people taken between 1441 and 1448. Subtracting those sold the month before the decree leaves 507.

4. "Doutra parte somos aquy vindos principalmẽte por quer lingua per queo O Iffãte nosso senhor possa saber novas desta terra que he cousa usque muyto deseia como todos bem sabees." Zurara, 88. This Berber speaker's initial conversations with the prince or court members likely occurred through an intermediary (a Berber-Arabic speaker).

5. Zurara, 92–93.

6. Zurara, 116.

7. *Romanus Pontifex*, January 6, 1454.

8. Zurara, 87. "Que serya bem jrmos a ella [Arguin island] e podera seer que nosso senhor Jhesu christo que sempre ajuda os que bem trabalham ordenara de auernos della algũa lingua." Zurara, 90. The concern with compensating *owners* is noted: "[C]adauno deli nostri nauilij haueano trucimanj negri portadi di portogallo i qual trucimanj sono schiauj negri vendudi per quel signor de Senega . . . i qual trucimanj se fesseno christianj in portogallo e Impresseno ben la língua spagnola I aueuemo habuti da suo misere per stipendio e soldi de darli una testa per vno a cernir in tuto nostro monte." Cadamosto, 52–53.

9. "Per suo fadiga dela trucimania e dando chadaun de questi truciman a suo misere 4 schiauj lor li lassano franchi e cossi per questo mezo molti schiauj son fati franchi dapoi per questo mezo dela trucimania" (By giving their owners four enslaved people [to replace themselves] they would be freed. By this means, many enslaved people became free, now and later, by using translation). Cadamosto, 53.

10. "Casting lots as to whose slaves would be sent to land first." It fell to the Genoese, whose slave was murdered during the first contact. Cadamosto, 53.

11. When sending a small armed ship ("barche al adsaltare") up into the Gambia, Cadamosto stated that "nuj eremo li conduti per uoler tratar con el paexe bona paxe" [*sic*]. Cadamosto, 54.

12. Wolof was widely spoken in Senegal from the mouth of the Senegal River south to Dakar. Mandingo appeared somewhere south of Dakar. Valentin Vydrin, "On the Problem of the Proto-Mande Homeland," *Journal of Language Relationship* 1 (2009): 107–42, esp. 115; Valentin Vydrin et al., *Mandé Language Family of West Africa: Location and Genetic Classification* (Dallas, TX: SIL International, 2000); Paul H. Hair, "An Ethnolinguistic Inventory of the

Upper Guinea Coast before 1700," *The Journal of African History* 8, no. 2 (1967): 247–68. The Bak group contained three unique languages: Balanta-Kentohe, Papel, and Manjak, some all spoken on a single island. Cadamosto, 73.

13. Cadamosto's account, written three years after Prince Henry's death, declares that King Afonso V sent both ships. Pacheco Pereira's, composed forty-seven years later, credits Henry with sending the first of the caravels, perhaps following Rui de Pina's official *Chronica*, composed at least thirty years after Henry's death (i.e., between 1490 and 1504). However, both may be correct if Henry ordered the first ship but did not survive to see its departure, leaving the actual departures to be carried out under Afonso V. Damião Peres attempted to resolve the confusion by suggesting an earlier journey as far as Sierra Leone, and a second farther south. In either case, King Afonso composed the orders for an initial journey. Cadamosto, 73; Rui de Pina, *Chronica d'el-rei D. Affonso V*, 3 vols. (Lisbon: Escriptorio, 1907), vol. 3, 144.

14. Cadamosto, 79. Little is known about this captain, except for what is found in a secondhand report to the Venetian navigator Alvise Cadamosto.

15. This deduction comes from two sources: the pattern of using slave interpreters on previous voyages and the detailed description of the interior of homes. Portuguese men would not have entered such small, enclosed places on foreign soil, especially without understanding the language. However, they would risk enslaved people's lives on such a venture. Cadamosto, 73–74.

16. Cadamosto mentions the expedition as passing by the Rio Besegue, named for the ruler who resided at its mouth, but does not mention anchoring there, apparently because contact had already been made with this community. The modern identification of the Besegue is uncertain, since it might have been one of three rivers: the Cacine, Konponi, or Nunez.

17. Cape Verga was mistakenly called Cape Sagres by Cadamosto's informant. This unnamed individual used "Verga" for Yongonsal, the only headland between the Bijagos archipelago and a diamond-shaped mountain (modern Verga). This retelling is further confused since the islands lie neither off Yongonsal nor off present-day Cape Verga but off the peninsula of Kaloum, the site of modern Conraky. The second language incomprehensible to Mandé speakers was likely Nalu, now nearly extinct. Cadamosto, 75–76.

18. The voyagers' mistaken assumption that they had reached a river likely stemmed from the peninsula's length on the bay's southern side. Cadamosto's informant wildly exaggerated the length of the Kaloum peninsula, claiming that it was sixty Italian miles long. Élisée Reclus, "Peuples des rivières du Sud," in *Nouvelle géographie universelle: la terre et les hommes*, vol. 12 (Paris: Hachette, 1876), 340; Cadamosto, 75–76.

19. "Seis São as ilhas dos ídolos (nome que tanto as cartas inglesas como as francesas adulteraram para o de ilhas de Loss)." Alexandre Magno de Castilho, *Descripção e roteiro da costa occidental de Africa*, 210.

20. Major rivers include the Pongo, Mellacorée, the junction of the Konkouré and Dubréka, and the Great and Little Scarcies. Without more precise descriptions, the exact locations of the Saint Vincent and Green Rivers cannot be determined with certainty.

21. Surf pounds the Moa River. The most likely candidate for the Smoky River is the Saint Paul River. Another contender is the Lofa River nineteen miles farther south. Cadamosto, 78.

22. "In questa parte vedeno molti fuochi quella prima note super li albori e por la spiazze fatti dalli Negri." Cadamosto, 78. Several years later, contact was made with the residents of the Saint Paul River just north of the cape, but reportedly the inhabitants had only small quantities of gold to trade.

23. Cadamosto, 78.

24. *The 1856 British African Pilot*, vol. 1, 104, describes the area south of Cape Mesurado as "nine miles sandy beach backed by forest."

25. Cadamosto, 79.

26. Cadamosto, 79. Although Afonso's exact words are lost, the repetition of the nouns for "peace," "love," and "force" in this sentence was common in official edicts.

27. "Per adimpir el comandamento del suo signor re . . . de menarli algun de quelli negri de paexe o per amor o per forza per poder intender de luj con tempo o per via de trucimania da molti altri negri che se troua in portagallo o per spatio de tempo che lo Imparasse a parlar . . . conduxeno a portagallo." The second goal "alcune nuoue de quelli so paesi di negri per questa cason passar piu oltra quello." Cadamosto, 79.

28. Cadamosto, 79.

29. This interest in Indigenous languages began informally before becoming a formal policy. In 1494, Hieronymus Münzer observed: "Habet item rex nigros . . . de vario idiomate, qui linguam portugalensem sciunt, quia várias linguas habent, et bis interpretibus usus quasi totam Aethiopiam superambulat." Hieronymus Münzer, "Do descobrimento da África marítima e ocidental, isto é, da Guiné pelo infante D Henrique de Portugal in Jerónimo Munzer," in *As Viagens dos descobrimentos*, ed. José Manuel Garcia (Lisbon: Editorial Presença, 1983), 357.

30. In his instructions for the expedition to the Anchorage of the Horses, Prince Henry expressed interest in learning any language spoken along the coast, declaring that any intelligence gained "would be no small thing." From this encounter forward, Portuguese voyagers usually included at least one translator, showing the sustained importance of Indigenous languages. Zurara, 55–58; Münzer, "Do descobrimento da África marítima e ocidental," 55–71, 67. The last fifteenth-century journey reporting the abduction of numerous natives to serve as interpreters occurred during Diogo Gomes's excursion to twenty-two degrees south of the equator.

31. The first such individual was Joao Fernandes, a squire of Prince Henry's court, who asked to stay during the three-caravel trip to Rio Douro in 1444, and who remained there for seven months. Zurara, 124, 139.

32. Antonio de Nebrija, *Gramática Castellana* (Halle, Belgium: Max Niemeyer, 1909), fol. 1.

33. Alexander George Findlay, *A Directory for the North Atlantic Ocean: Comprising Instructions General and Particular for Its Navigation*, 16th edition (London: Imray, Laurie, Norie & Wilson, 1918), 162.

6. A PAINTED SHIP UPON A PAINTED OCEAN

1. Samuel Taylor Coleridge, *The Rime of the Ancient Mariner* (Boston: Estes and Lauriat, 1884).

2. Roger Graham Barry and Richard J. Chorley, *Atmosphere, Weather, and Climate* (London: Routledge, 1992). The Intertropical Convergence Zone is commonly abbreviated ITCZ, pronounced "itch."

3. The Indian Ocean monsoon was mentioned as early as 312 BCE. Bruce A. Warren, "Ancient and Medieval Records of the Monsoon Winds and Currents of the Indian Ocean," in *Monsoons*, ed. Jay S. Fein and Pamela L. Stephens, 137–58 (New York: John Wiley & Sons, 1987), 140–41. "Recent theoretical advances and observational analyses support a novel view of monsoons as localized seasonal migrations of the tropical convergence zone." Ruth Geen et al., "Monsoons, ITCZs, and the Concept of the Global Monsoon," *Reviews of Geophysics* 58, no. 4 (2020). The northern branch of the Indian Ocean ITCZ lies at fifteen degrees north; the southern branch, slightly south of the equator, lacks any significant commerce. Pang-chi Hsu and Tim Li, "Is 'Rich-Get-Richer' Valid for the Indian Ocean and Atlantic ITCZ?," *Geophysical Research Letters* 39, no. 13 (2012). During summer in the Northern Hemisphere, the ITCZ shifts northward, bringing moist warm air from the ocean, helping to establish the Southwest Monsoon over the Arabian Sea, leading to heavy rains in India. In the South China Sea, from September through February, the currents move southwestward between 110 and 120 degrees east. During spring (March–May), they tend to move eastward from Indonesia and Borneo toward China and Vietnam. The winds move northeastward in June, July, and August, allowing for a return trip. Huei-Ying Kuo, "Charting China in the Thirteenth-Century World: The First English Translation of *Zhu fan zhi* and Its Recipients in China in the 1930s," in *Knowledge in Translation: Global Patterns of Scientific Exchange, 1000–1800 CE*, ed. Patrick Manning and Abigail Owen, 93–116 (Pittsburgh: University of Pittsburgh Press, 2018); Daji Huang et al., "Alongshore and Cross-Shore Circulations and Their Response to Winter Monsoon in the Western East China Sea," *Deep Sea Research Part II: Topical Studies in Oceanography* 124 (2016): 6–18; Bin Wang et al., "Multi-Scale Climate Variability of the South China Sea Monsoon: A Review," *Dynamics of Atmospheres and Oceans* 47, nos. 1–3 (2009): 15–37, esp. fig. 3; Lynda Norene Shaffer, *Maritime Southeast Asia to 1500* (London: M.E. Sharpe, 1996); Kenneth R. Hall, *A History of Early Southeast Asia: Maritime Trade and Societal Development, 100–1500* (Lanham, MD: Rowman & Littlefield, 2011); Roderich Ptak, "From Quanzhou to the Sulu Zone and Beyond: Questions Related to the Early Fourteenth Century," *Journal of Southeast Asian Studies* 29, no. 2 (1998): 269–94.

4. Only one monsoonal pattern exists in the easternmost corner of the Gulf of Guinea. Marion Leduc-Leballeur et al., "Air–Sea Interaction in the Gulf of Guinea at Intraseasonal Time-Scales: Wind Bursts and Coastal Precipitation in Boreal Spring," *Quarterly Journal of the Royal Meteorological Society* 139, no. 671 (2013).

5. The Western Pacific lies east of −149 degrees longitude. Heading to Hawaii, canoes could start northwest, then trail the eastward current until they

were due south of their destination. Returning to the Society Islands, they could depart to the southwest and follow the same current until just north of their objective. Jonathan K.P. Shonk et al., "Identifying Causes of Western Pacific ITCZ Drift in ECMWF System 4 Hindcasts," *Climate Dynamics* 50 (2018): 939–54. Polynesians did not cross the equator in the Western Pacific, which lies on the opposite side of the Pacific's geographic center (close to Rapa Nui at 48°52.6′S, 123°23.6′W). Athanasios Koutavas and Jean Lynch-Stieglitz, "Variability of the Marine ITCZ over the Eastern Pacific during the Past 30,000 Years," in *The Hadley Circulation: Present, Past and Future*, ed. Henry F. Diaz and Raymond S. Bradley, vol. 21 of *Advances in Global Change Research* (Dordrecht, The Netherlands: Springer, 2004). Hence, nations crash their spacecraft debris there. Vito De Lucia and Viviana Iavicoli, "From Outer Space to Ocean Depths: The Spacecraft Cemetery and the Protection of the Marine Environment in Areas Beyond National Jurisdiction," *California Western International Law Journal* 49 (2019): 34–90; Shonk et al., "Identifying Causes of Western Pacific ITCZ Drift." The disappearance of the doldrums during April in the Pacific, along with the steady eastward flow of the Equatorial Countercurrent, helped Polynesian vessels navigate the Pacific doldrums. Ben R. Finney, *Voyage of Rediscovery: A Cultural Odyssey through Polynesia* (Berkeley: University of California Press, 1994), 66, 203–4, 236–38.

6. Capillary or gravity capillary waves. For gravity waves, see Adrian E. Gill, "Gravity Wave," in *Glossary of Meteorology*, ed. Todd Glickman (Boston: American Meteorological Society, 2000).

7. The plant is rhizomatic, but its shoots produce shrub-sized growths.

8. Shang-Ping Xie and James A. Carton, "Tropical Atlantic Variability: Patterns, Mechanisms, and Impacts," in *Earth Climate: The Ocean-Atmosphere Interaction*, ed. Chunzai Wang, Shang-Ping Xie, and James A. Carton (Washington, DC: American Geophysical Union, 2004).

9. The Sahara's boundaries are customarily given as between eleven and twenty-three degrees north. The exact distance to the southern border of the Sahara is 1,558 land miles (2,556 kilometers) or 1,380 nautical miles. The southern boundaries of the thermal equator in West Africa lie 760 land miles (1,222 kilometers) or 660 nautical miles north of the geographic.

10. On the southeastern side of the African continent, the Drakensberg or Dragon Mountains push the heat equator dramatically far south to Antananarivo, the capital of Madagascar, nineteen degrees past the equator. Antananarivo at 18.916667, 47.516667 lies 1,174 land miles and over a thousand nautical miles south of the geographic equator.

11. Landforms, processes, and impacts of wind are called "Aeolian." The ITCZ drifts slowly northward across the open waters of the Indian Ocean, crossing the geographic equator near Indonesia and then trending southward toward the Cook and Solomon Islands in the Pacific. Amin Dezfuli, "Climate of Western and Central Equatorial Africa," in *Oxford Research Encyclopedias Africa* (Oxford: Oxford University Press, 2017).

12. Aristotle [384–322 BCE], *Meteorology*, bk. 2, pt. 5. Repeated by Virgil [70–19 BCE]: "quinque tenent caelum zonae: quarum una corusco semper sole

rubens et torrida semper ab igni." Virgil, *Georgics*, bk. 1, 232–34. See also Strabo [64 BCE–21 CE], *Geographica*, bk. 2, chap. 3, sec. 1; Ovid, *Metamorphoses*, lines 45–50. This view prevailed but was not universal.

13. The Tropic of Cancer lies at 23.436806, Capricorn at −23.436278.

14. Hieronymus Münzer, "Do descobrimento da África marítima e ocidental, isto é, da Guiné pelo infante D Henrique de Portugal in Jerónimo Munzer," in *As Viagens dos descobrimentos*, ed. José Manuel Garcia, 55–71 (Lisbon: Editorial Presença, 1983 [1495]), 55–71, 65.

15. On an overcast or even partly hazy day, the clouds absorb nearly 20 percent of the sun's heat.

16. William Morris Davis and William Henry Snyder, *Physical Geography* (Boston: Ginn, 1894) (uses *heat equator* thirteen times).

17. The kettle takes so long to boil because an enormous amount of heat is needed to break the hydrogen bonds (separating H_2 from H_2O).

18. John Marshall et al., "The Ocean's Role in Setting the Mean Position of the Inter-Tropical Convergence Zone," *Climate Dynamics* 42 (2014): 1967–79. "Due to the Atlantic ocean's meridional overturning circulation, however, the ocean carries significantly more heat northwards across the equator (order 0.4 PW) . . . [A]tmospheric heat transport is [carried] southwards across the equator to compensate (0.2 PW southwards), resulting in the ITCZ being displaced north of the equator." Brian Green et al., "Twentieth-Century Correlations between Extratropical SST Variability and ITCZ Shifts," *Geophysical Research Letters* 44 (2017): 9039–47.

19. As discussed elsewhere in this chapter, commercial voyages to the area south of the Sinoe River were subject to the doldrums.

20. Between 150 and 200 miles offshore, strong seasonal reversing winds and currents accompanied the episodic calms. Alexander George Findlay, *A Directory for the North Atlantic Ocean: Comprising Instructions General and Particular for Its Navigation*, 16th edition (London: Imray, Laurie, Norie & Wilson, 1918), 157.

21. The Valentim Fernandes manuscript repeated the warning about the giant rocks "made as islands stuck to the ground": "Sabe q do ryo do Jũco [south] ao Resgate do Soeyro tudo sõ pedras ao lõgo do mar feytas como ylheos pegados na terra. . . . E do ryo do Jũco ao Resgate do Soeyro ha ilij legoas e tudo he penedia como te digo." *CVF*, 220. Fernandes, descriptions of the rocks match contemporary ones. *Africa Pilot*, vol. 1, chap. 8, secs. 92, 196, 198; vol. 1, chap. 9, sec. 11.

22. The exact latitudes of the coastal boundaries correspond to the area from the Pongo to the Sinoe River. *Africa Pilot*, vol. 1, chap 8, sec. 186.

23. Cape Mesurado to Junk/Farmington/Garbage River: twenty-eight miles east-southeast, with entrance at 6.104167 N, 10.39944 W. The "sandy beach" quote can be found in countless English-language sailing guides, e.g., U.S. Hydrographic Office, *Sailing Directions for the West Coasts of Spain, Portugal, and Northwest* (Washington, DC, 1908), 315; *Africa Pilot*, vol. 1, chap. 8, secs. 187, 192, 224.

24. "Este Rio do Junco tem hum Ilheo na boca e aqui há também ouro em pouca quantidade e pelo mesmo modo malagueta." *Esmeraldo* (Évora), fol.

62v. "Sabe q~ ho ryo do Jũco tem estes synaes, huã serra grãde em cyma de sy. E a maior serra fica da bamda do loeste. E da bãda de leste tẽ tres ou quatro ylheos pequenos. E da bãda de oeste huã aruore grossa redonda soo." *CVF*, 220. Since then, the villages on the south side of the river have disappeared and the hill only harbors chimpanzees.

25. "E dally [ryo do Jũco] auãte acharas almadias se a terra for segura. . . . E do ryo do Jũco ao Resgate do Soeyro . . . tudo he penedia como te digo." *CVF*, 220, which lists many forests between the Rio do Jũco and Rio dos Cestos. *Esmeraldo* (Évora), fol. 64, has the Resgate de Genoese after Cestos. *Africa Pilot*, vol. 1, chap. 8, 192–196. *CVF* (220) has many forests between the Rio do Jũco and Rio dos Cestos. *Esmeraldo* (Lisbon), fol. 52v, has the Resgate de Genoese after Cestos.

26. *Aframomum melegueta* is native to swampy regions of the West African coast. Pods measure five to seven centimeters (2.0–2.8 inches). Solomon Umukoro and Aderemi Caleb Aladeokin, "Therapeutic Effects of Grains of Paradise (*Aframomum melegueta*) Seeds," in *Nuts and Seeds in Health and Disease Prevention*, ed. Victor R. Preedy et al. (San Francisco: Academic Press, 2011), 535–43; K.V. Peter and M.R. Shylaja, "Appendix 1 (ISO List of Plant Species)," in *Handbook of Herbs and Spices*, ed. K.V. Peter, 2nd edition (Oxford: Woodhead, 2012), vol. 1; A.M. Van Harten, "Malagueta-peper-een literatuuronderzoek betreffende de historische Grains of Paradise," *Landbouwkundig Tijdschrift* 78 (1966): 177–82. André Almada (flourished ca. 1570) describes it as a folk remedy for stomach and intestinal pain. André Donelha and Avelino Teixeira Mota, *Descrição da Serra Leoa e dos rios de Guiné do Cabo Verde (1625)* (Lisbon: Junta de Investigações Científicas do Ultramar, 1977), 68. Usodimare had previously seen Malagueta peppers in 1455. Antoniotto Usodimare, "Carta de Antoniotto Usodimare," in *Viagens dos descobrimentos*, ed. José M. Garcia (Lisbon: Editorial Presença, 1983), 139–44. The name began to appear on maps, starting with the increased popularity of the spice in Northern Europe. Where this coast began varied. For example, in 1822, the British admiralty defined the Grain Coast as extending from Cape Mount to Cape Palmas; in 1849, the French used the identical boundaries; but by 1885, the British had moved the northern boundary up to the Shoals of St. Ann.

A Dutch governor later commented critically on the Portuguese designation "Grain or Pepper Coast," claiming it came from a local word meaning a welcome or greeting, *quaqua-kust*. According to a note of Governor Valckenburgh in 1659, "Grein of Peperkust . . . komt ook voor onder den naam Quaqua-kust, volgens eene nota van den Gouverneur Valckenburgh Anno 1659, naar een woord dat aldaar een begroeting inhield." Pieter de Marees and Samuel Pierre L'Honoré Naber, *Beschryvinghe ende historische verhael van het Gout koninekrijek van Gunea anders de Gout Custe de Mina genaemt liggende in het deel van Africa* (The Hague: M. Nijhoff, 1912 [1602]), 15.

27. *Esmeraldo* (Évora), fol. 51v, spells it "Reguebos"; *Esmeraldo* (Lisbon), "Zebuebos."

28. Mouth River Cess (5.4458284, −9.5829660), Monrovia 5.313333, −10.801389 (Haversine calculation). *Esmeraldo* (Évora), 300. La Fosse described the plant as "le fruit qu'elle produit est gros et rouge comme une

pomme rouge de rubis, et . . . cette graine les habitants font des colis empaquetes de leurs propres feuilles, liés de leurs propres tiges et pesant 10, 20 ou 30 livres; bien formés en carrés, ils sont chose fort plaisante à voir." Eustache de La Fosse et al., *Voyage d'Eustache Delafosse sur la côte de Guinée, au Portugal et en Espagne (1479–1481)* (Paris: Editions Chandeigne, 1992), 24–25, 27.

29. "[E] esta trazem em huns cestos o que em toda há outra costa honde há a dita malagueta nom custumam trazer e para segurança do Navio e gente." *Esmeraldo* (Évora), fol. 63. *Esmeraldo* has the mouth of Cestos at 5.5 degrees north.

30. "New Cess River—coast to southeast" was fronted by cliffs forty-six meters high near the river entrance, and large trading vessels could anchor off the river. *Africa Pilot*, vol. 1, chap. 8, secs. 167, 200.

31. His name was Soeiro da Costa, hence the place name "Resgate de Soeiro," or sometimes "Resgate do Genoves." In this text, *resgate* was employed for both the trade in enslaved people and Malagueta. *Africa Pilot*, vol. 1, chap. 8, 18. Andrea Massing identifies Soeiro as an aged former official from Lagos (337), but *Esmeraldo* (Évora) states that he was a Genoese sailor: "marinheiro genoes." Andrea Massing, "Mapping the Malagueta Coast: A History of the Lower Guinea Coast, 1460–1510 Through Portuguese Maps and Accounts," *History in Africa* 36 (2009): 331–65.

32. Valentim Fernandes and João de Castro place the Resgate de Soeiro before the Cestos River. Still, both copies of *Esmeraldo* have it after Cestos, as do six Iberian maps made between 1471 and 1510. Fernandes describes the place as follows: "Ho Resgate do Soeyro tem estes synaes da bamda do hoeste tem huã serra pequena no sertão q faz quatro mõtes. E como fores tãto auãte grãdes soos q~ parecẽ pinheyros." *Esmeraldo* (Évora), 64; *Esmeraldo* (Lisbon), fol. 52v; *CVF*, 22. The maps include Jorge Aguiar (1492); Colon, Cantino four-finger map (undated); and British Library, Edgerton ms 73 f. 64, Modena C.G.A.5 (undated). Archaeology could likely identify the location of the Resgate de Soeiro.

33. *Esmeraldo* (Évora), fol. 64v.

34. See note 26 in this chapter.

35. *Esmeraldo* (Évora), fol. 64v–65; *Esmeraldo* (Lisbon), fol. 63. In Fernandes's manuscript it is spelled *guey*, pronounced in Portuguese like *ewe* in American English. The text adds another characteristic of this group: They were only ever known by a single name. Although they are thought to have originated farther east, their presence here suggests earlier migration. *CVF*, fol. 64v.

36. The dangers lay in succession: Cestos Reef, Schooner, Sunk Rocks, Rock Cess Point, Diabilitos Reef, and Mutine Rock. The original name of the place where enslaved people were held may have differed. However, a tiny hamlet named Kei Village is at the outlet of the Bioni River. The Évora and Lisbon manuscripts mention that the people residing on the coast were highly skilled fishermen who traveled considerable distances to sea. *Esmeraldo* (Évora), fol. 63v; *Esmeraldo* (Lisbon), fol. 52–52v. For the likely sources of enslaved people, see https://lhsliberia.com/wp-content/uploads/Appendix-16-09-Sehnkwehn-River.pdf. *Africa Pilot*, vol. 1, chap. 8, sec. 195. The town on the Bioni was Kei. "Os negros de toda esta terra trazem ha malagueta a Resgatar ao navios

nas almadia em que vaaõ a pescar ao mar. Andam nuus e nom som cicursisos." *Esmeraldo* (Évora), fol. 65. During the eighteenth century, British vessels began to recruit these Kru men as deckhands and sometimes sailors aboard British vessels.

37. *Esmeraldo* (Évora), fol. 64v; *Esmeraldo* (Lisbon), fol. 53. The locations of the Resgate de Soeiro, Rio Sao Vicente, and Praia dos Escravos are best guesses, since the historical data do not align with any later map or nautical guide, including those in seventeenth-century Dutch. The sources do not identify which African society traded with the Portuguese on this coast.

38. *Esmeraldo* (Évora), fol. 64v–65; *Esmeraldo* (Lisbon), fol. 63; *CVF*, fol. 64v. Later, the center of trade in enslaved people and Malagueta peppers moved to the north, where both were more available.

39. *Esmeraldo* (Évora), fol. 64.

40. "[D]aRedor destes Ilheos ha mtos baixos de pedra mto perigosos e maaos e delles paresem sobre augua e outros nom/l e q.mnesta Costa ouuer de andar com nauio grd.[e] de 80 ou cem tonees pouse pellas 35 braças e estaraa huma grde. Leguoa e mea da terra." *Esmeraldo* (Lisbon), fol. 52. Virtually identical warnings appear three hundred years later. "The reef and all these rocks are the more dangerous because they are all so steep that the lead [plumb line] can give little or no warning. . . . A vessel passing along the coast may avoid them all, by keeping outside of 14 fathoms." UK Admiralty, *Sailing Directions for the Western Coast of Africa*, vol. 1, 1856, 108.

7. GOLD AT LAST

1. Studies of this practice appear in several publications: Kide Yemane et al., "Assessing the Main Challenges of Artisanal and Small-Scale Mining Operations in Africa: A Simple Overview," *Geominetech: The Indian Mineral Industry Journal* 7, no. 2 (2020); J. Mantey et al., "Operational Dynamics of 'Galamsey' within Eleven Selected Districts of Western Region of Ghana," *Journal of Mining & Environment* 8, no. 1 (2017): 11–34; Frederick Owusu-Nimo et al., "Spatial Distribution Patterns of Illegal Artisanal Small Scale Gold Mining (*Galamsey*) Operations in Ghana: A Focus on the Western Region," *Heliyon* 4, no. 2 (2018): e00534; Anthony Aubynn, "Sustainable Solution or a Marriage of Inconvenience? The Coexistence of Large-Scale and Artisanal and Small-Scale Mining on the Abosso Goldfields Concession in Western Ghana," *Resources Policy* 34, no. 1 (2009): 64–70; BBC News, "Women Die in Ghana Mine Collapse," November 12, 2009. For a NASA photograph of the extent of artisanal mines, see Chrissy Sexton, "Satellite Imagery Exposes Illegal Gold Mining in Ghana," June 8, 2021, https://www.earth.com/image/satellite-imagery-exposes-illegal-gold-mining-in-ghana/.

2. *Esmeraldo* (Évora), fol. 53v, bk. 3, chap. 4.

3. *CVF*, 225; *Esmeraldo* (Évora), fol. 53v, bk. 3, chap. 4; UK Hydrographic Office, *The African Pilot*, 1856; U.S. Hydrographic Office, *Sailing Directions for the West Coasts of Spain, Portugal, and Northwest* (1908), 58. One mile per hour on a twelve-hour day at 0.95 knots.

4. *CVF*, 225 (emphasis added).

5. For the current's push from Elmina eastward, see map 9. The current's most vigorous push extended east from Cape Palmas two hundred miles short of Fernando Po (Bioko). In the mid-nineteenth century, Castilho wrote that "comprovado pela experiencia, que durante a Estação das chuvas, de Maio a Outubro a Corrente de Guiné se dirige quasi constantemente a E. com grande velocidade." Alexandre Magno de Castilho, *Descripção e roteiro da costa occidental de Africa desde o cabo de Espartel até o das Agulhas* (Lisbon: Imprensa Nacional, 1866), vol. 2, chap. 10.

6. *CVF*, 129. Fernandes adds that the island was found by a caravel of Fernão de Mello, who became lost on his way to São Tomé.

7. The uninhabited island became São Tomé, 137.36 nautical miles from Fernando Po. It was discovered in 1492 and first settled on the feast of John the Baptist in 1503. *CVF*, 122. From the midpoint of each island, the distance between islands reaches 150.1 nautical miles (all Haversine calculations), a distance Fernandes reported as thirty leagues. *CVF*, 129. The Pico de São Tomé rises 6,640 feet (2,024 meters).

8. Antonio Lopes da Costa Almeida, *Roteiro geral dos mares, costas, ilhas, e baixos reconhecidos no globo, parte IV: Costa d'Africa de cabo de Espartel até cabo de Boa Esperança, com as ilhas adjacentes* (Lisbon: Typografia de Academia, 1835–1849, 4 vols.), 297. "Assim entre os Meridianos de 2° O. e 16° E. de Lisboa na distancia de 1000 m. se encontra a notavel anomalia de existirem duas Correntes com consideravel velocidade, conduzindo suas agoas em direcção parallela e diametralmente oposta."

9. A warmer Atlantic increased the eddies that formed between Cape Palmas and Cape Three Points. D. Binet, "Climate and Pelagic Fisheries in the Canary and Guinea Currents 1964–1993: The Role of Trade Winds and the Southern Oscillation," *Oceanologica Acta* 20, no. 1 (1997): 177–90; Alexander B. Polonsky et al., "A Study of the Equatorial Atlantic Currents Using Historical Moored Data," *Deutsche Hydrografische Zeitschrift* 45, no. 4 (1993): 219–54; *CVF* (Lisbon), fol. 307.

10. U.S. Hydrographic Office, *Sailing Directions* (Washington, DC, 1916), 96. Eddies also formed around both Cape Palmas and Three Points. Binet, "Climate and Pelagic Fisheries," 177–90.

11. A.A. Boahn, "The States and Cultures of the Lower Guinean Coast," in *General History of Africa*, abridged edition, vol. 5, *Africa from the Sixteenth to the Eighteenth Century*, ed. Bethwell A. Ogot (Paris: UNESCO, 1992), 399–433.

12. Of the Dinkira, Bosman wrote: "The Dinkiras possessed a great wealth of Gold . . . [and had] three lands under their power named Wassa, Encasse, and Jusser where gold (but not much) was found." Willem Bosman [1672–1703], *A New and Accurate Description of the Coast of Guinea, Divided into the Gold, the Enslaved Person, and the Ivory Coasts . . . by Willem Bosman, Chief Factor for the Dutch at the Cafe of St. George d'Elmina. And Now Faithfully Done Into English* (London: Printed for James Knapton . . . and Dan Midwinter, 1705), 74.

13. Sam Piers, "The Eguafo Kingdom: Investigating Complexity in Southern Ghana" (PhD diss., Syracuse University, 2007), 25; Christopher R. DeCorse, "Coastal Ghana in the First and Second Millennia AD: Change in Settlement

Patterns, Subsistence and Technology," *Journal des africanistes* 75, no. 2 (2025): 43–54.

14. Allusions appear to two pilots who worked for Lisbon merchant Fernão Gomes: Pêro Escobar and Álvaro Esteves, each of whom has been suggested as possibly heading the first expedition to Mina. *Esmeraldo* (Évora), fol. 68.

15. *Africa Pilot*, vol. 1, chap. 10, secs. 16, 20, 21, 71. The 3.5-mile-wide Takoradi Bay lies around six miles east of the cape. Elmina anchorage: *Africa Pilot*, vol. 1, chap. 10, sec. 88. Although no records survive of the actual protocol the Portuguese followed, this account describes their standard approach since the 1450s, following the disastrous encounter in the Sine-Saloum.

16. Avelino Teixeira da Mota's exact words: "The caravel triumphed over the caravan in the trade of . . . slaves . . . [and] gold." Avelino Teixeira da Mota, *Some Aspects of Portuguese Colonization and Sea Trade in West Africa in the 15th and 16th Centuries* (Bloomington: African Studies Program, Indiana University, 1978), 4.

17. The amicable relations between the two towns led the Portuguese to initially christen the place Twin Villages, a name that referred to the existing native settlements. However, they added another name, Saint George of the Mine (São Jorge Da Mina). Although the year of their arrival remains uncertain, this name suggests that the date was April 23, the feast of Saint George. Portuguese maps retained both names for several decades: "L'aldeia das duas partes (ce nom lui a été donné parce qu'il s'agit de deux villages, distants l'un de l'autre d'un trait d'arc)." La Fosse, 27; Manoel Fuigueiredo and Biblioteca Complutense [Alcalá de Henares], *Hidrographia, Exame De Pilotos No Qual Se Contem As Regras Que Todo Piloto Deue Guardar Em Suas Nauegaço[n]es* (Lisbon: Vicente Aluarez, 1625), 57.

18. The Portuguese also constructed a fort on the western side of Cape Three Points named Axim but abandoned it because it was slower and more dangerous than the route to Mina. Bosman, *A New and Accurate Description of the Coast of Guinea*, 73, 80–81. There may also have been more malaria present. *Esmeraldo* (Évora), 117.

19. A summary of many papers discussing issues of obtaining bullion appears in J.H. Munro, "Medieval Monetary Problems: Bimetallism and Bullionism," *The Journal of Economic History* 43, no. 1 (1983): 294–98; Martin Allen, "Medieval Merchants and the English Mints and Exchanges," in *Medieval Merchants and Money: Essays in Honour of James L. Bolton*, ed. Matthew Davies and Martin Allen (London: University of London Press, 2016), 197–212; John Day, "The Great Bullion Famine of the Fifteenth Century," *Past & Present* 79, no. 1 (1978): 3–54; Boaz Shoshan, "Exchange-Rate Policies in Fifteenth-Century Egypt," *Journal of the Economic and Social History of the Orient* 29, no. 1 (1986): 28–51.

20. John Day characterizes this shortage as a "famine," an incredibly wrongheaded and insulting use of the term. People can neither eat nor drink gold, as the legend of King Midas reminds us. Day, "The Great Bullion Famine," 3–54. A gold deficit mattered only to people with plenty of food and water. For a detailed monetary history of gold and silver in Portugal and Europe, see Vitorino Magalhães Godinho, *Os descobrimentos e a economia mundial*, 2nd edition, 3

vols. (Lisbon: Editorial Presença, 1981), vol. 1, 117–38. Silver shortages also abounded because of the exhaustion of Northern European mines.

21. King Afonso V of Portugal invaded Castile on May 10, 1475, and married Joanna in Plasencia fifteen days later. Joanna signed her letters "La Reina" (The Queen) until she died. She would become known in Portugal as "a Excelente Senhora" (the Excellent Lady). See Isabel Vaz de Freitas, *D. Joana, A Excelente Senhora: 1462–1530* (Vila do Conde, Portugal: Quidnovi, 2011); Isabel Vaz de Freitas, "D. Joana: Uma mulher entre Portugal e Castela," in *En La Europa medieval, mujers com historia, mujers de leyenda Siglos XIII–XVI*, ed. Manuel García Fernández (Seville, Spain: Universidad de Sevilla, 2019); "Capitulações dos reis de Castela para o contrato de casamento de D. Afonso V (1470–1472)," trans. Diana Martins, in *Fragmenta Histórica—História, Paleografia e Diplomática*, ed. João José Alves Dias (Lisbon: Centro de Estudos Históricos, 2018), vol. 6, 53–55; Susannah Humble Ferreira, "Juana La Beltraneja, Dynastic Fears, and Threats of Marriage (1475–1506)," *Renaissance and Reformation / Renaissance et Réforme* 43, no. 4 (2020): 79–100.

22. Isabel would only make such a claim knowing that her pilots could navigate safely to Mina and back. While a Spanish ship might have reached Elmina earlier, for Isabel to have received the mine's gold by May 1475, an Iberian vessel would have had to depart to poach the Elmina trade by the end of 1474 at the latest.

23. Spanish chroniclers invented many excuses for the defeat at Mina, from faulting the commanders for overstaying at Mina to blaming illness among the crew, exaggerating the size of the Portuguese fleet, and even claiming that the attack was a surprise (!)—in short, every pretext except the truth: that a superior naval power bested them. Hernando del Pulgar, *Crónica de los Reyes Católicos (1481–1490)*, ed. Juan de Mata Carriazo (Madrid: Espasa-Calpe, 1943); Alfonso Fernández de Palencia and Antonio Paz y Meliá, *Crónica de Enrique IV* (Madrid: Ediciones Atlas), vol. 4, bk. 25, chap. 5. King João II organized a permanent navy not long after this incident.

24. On March 6, 1480, in ratifying the Treaty of Alcáçovas, or Treaty of Toledo, Ferdinand and Isabella of Spain recognized Afonso V's African conquests, and he ceded the Canary Islands to Spain while retaining all the other Atlantic island chains.

25. Afonso V's sister Lenora married Frederick III, the Holy Roman Emperor. Maximilian I (Frederick's son) was the father of Charles V, Holy Roman Emperor (and Charles I of Castile and Aragon). Charles's maternal ancestry was from Queen Isabel.

26. French ships successfully attacked Mina in 1530 and 1531. João Cordeiro Cereira, "Resgate do ouro na Costa da Mina nos reinados de D. João III e D. Sebastião," *Studia: revista semestral* 50 (1991): 16. The first English ship led by Thomas Wyndham and a Portuguese pilot reached Guinea in 1553. At least two Flemish expeditions to Elmina failed. One wrecked on the rocks along the Pepper Coast, where the Ewe people (in 1490) recounted to a smugly satisfied Portuguese audience how they had eaten the Flemish crew. *CVF*, fol. 64v.

27. Eustache de La Fosse et al., *Voyage d'Eustache Delafosse sur la côte de Guinée, au Portugal et en Espagne (1479–1481)* (Paris: Editions Chandeigne, 1992).

28. La Fosse, 29.

29. La Fosse, 31; Barros, *Décadas* 1, bk. 3, chap. 3–4. Contrary to the usual understanding of Rio da Volta, Bosman describes "Rio Volta, probably so called by the Portuguese, by reason of its rapid Course and Reflux, discharging its Waters so violently into the Sea, that it is sometimes visible three or four Miles from the Shore. The extraordinary rapid Reflux into the Sea, continually carries great numbers of Trees with it." Bosman, *A New and Accurate Description of the Coast of Guinea*, 328, Letter 18. The exact location cannot be identified definitively, since the two surviving maps from the 1470s list two indecipherable towns past the Rio da Volta, the largest of which was Grand Popo. Since the exact boundaries of the Benin Empire in 1480 are open to debate, the polity and names of the initial traders remain unknown.

30. Barros, *Décadas* 1, bk. 3, chap. 3, describes the slave trade from the empire of Benin, where slaves were sold for double their acquisition cost to gold merchants who needed the labor. This practice continued during the reign of João II and Manuel. According to La Fosse, the Portuguese ships returned on Shrove day. La Fosse, 26–27, 33. Counting back from the Julian calendar date of Easter on April 2, Lent started on Wednesday, February 16, 1480. Thus, Portuguese ships arrived in Elmina harbor on January 6 and returned from Grand Popa precisely six weeks later. Since La Fosse's account fails to specify how long the Portuguese ships spent at Elmina until they departed, the maximum time for the voyage was six weeks.

31. La Fosse, 26–27. Each weight represented 3 sterling, making each sale worth 36–48 pounds sterling, but as Wyffels and Munro point out, exact equivalencies were hard to establish. Carlos Wyffels, "Notes sur les marcs monétaires utilisés en Flandre et Artois avant 1300," *Handelingen van het Genootschap voor Geschiedenis* 104, nos. 1–2 (1967): 66–87; Carlos Wyffels, "Contribution à l'histoire monétaire de Flandre au XIIIe siècle," *Revue belge de Philologie et d'Histoire Année* 45, no. 4 (1967): 1113–41; John Munro, "A Maze of Medieval Monetary Metrology: Determining Mint Weights in Flanders, France and England from the Economics of Counterfeiting, 1388–1469," *Journal of European Economic History* 29, no. 1 (2000): 173–99. The dangers and division of labor in the gold industry are discussed later in this chapter.

32. Olfert Dapper, *Naukeurige beschrijvinge der Afrikaensche gewesten van Egypt, Barbaryen, Libyen, Biledulgerid, Negroslant, Guinea, Ethiopiën, Abyssinie . . . : getrokken uit verscheyde hedendaegse lantbeschrijvers en geschriften van bereisde ondersoekers dier landen* (Amsterdam: Jacob van Meurs, 1686), 466.

33. Dapper, 466–67.

34. Pieter de Marees and Samuel Pierre L'Honoré Naber, *Beschryvinghe ende historische verhael van het Gout koninckrijck van Gunea anders de Gout Custe de Mina genaemt liggende in het deel van Africa* (The Hague: M. Nijhoff, 1912 [1602]), 195; Wilhelm Johan Mueller, *Die Africanische auff der Guineischen Gold-Cust gelegene Landschafft Fetu* (Hamburg, Germany: n.p., 1673).

35. Bosman, *A New and Accurate Description of the Coast of Guinea*, 79–80, Letter 6.

36. "U[we] E[xcellentie] weet wel beter en dat selfs tot de mijnen geen toegang hebben . . . 'It en geloof ook niet eens dat iemand ban ons ooit enige ban deselbe heeft gessen . . . want de Swarten dit als een Heyligdom aanmerkende sullen alle middelen aantwenden om 'er andere uyt houden." Willem Bosman et al., *Nauwkerige beschryving van de Guinese Goud-, Tand- en Slave-kust* (Utrecht: Anthony Schouten, 1704), 80.

37. "Negers wetender ons geen net denkbeeld ban te geben en ban ons komt'er niemand soo ber landwaards in dieshalben Mijn vriend zijt hier meede te vrede." Bosman et al., 80.

38. "Te weten de Koningen, wel bewaert worden: want ieder koning heeft zijne mijn-bergen, daer uit hy door zijn volk het gout laet delven." Dapper, *Naukeurige beschrijvinge*, 466. Nearly identical phrasing is found in Marees and L'Honoré Naber, *Beschryvinghe ende historische*, 194: "[N]aer het ondersoecken vande Negro scanickt qualijck ghedoen, deur de onsekerheydt die ick daer van hebbe: want ick achte datter luttel Negros zijn die inde zee plaetsen woonen, die de Mynen selver we ten, ende noyt en hebben by geweest, iae noch Portegues noch Neer lander selver, niemant by of ontrent en hebben geweest, dan dat de Mynen stil ghehouden worden, en de deur de eyghenaers wel worden bewaert: want ick hebbe verstaen van sommighe Negros dat elcken Coninck zijn Mijn-berghen heeft." The word *ruler* is the preferred modern translation of the seventeenth-century *koning*.

39. Identification by Gwendolyn Midlo Hall, "Africa and Africans in the African Diaspora: The Uses of Relational Databases," *American Historical Review* 115, no. 1 (2010–12): 136–50.

40. "Kasteel de la Mina, thans onder den naam van St. George del Mina of eenvoudig onder dien van DELMINA bekend." Dapper, *Naukeurige beschrijvinge*, 466–67. *Mina* is a feminine noun in both Portuguese and Spanish but gender-neutral *de mijn* in Dutch. During the seventeenth century, the Dutch changed *da Mina* to *Del Mina*, an easier pronunciation in Dutch. Eventually the *d* was dropped, leaving only *Elmina*.

41. António de Almeida Mendes, "Les réseaux de la traite ibérique dans l'Atlantique nord (1440–1640)," *Annales. Histoire, Sciences Sociales* 63, no. 4. (2008): 739–68.

42. Commercial fishermen, lumberjacks, farmworkers, and underground miners all die at far higher rates than fire or police personnel. They join roofers, construction workers, iron and steel workers, collectors of refuse and recyclable materials, and truck drivers among the deadliest occupations in the United States. U.S. Department of Labor, Bureau of Labor Statistics, *Injuries, Illnesses, Fatalities*, https://www.bls.gov/iif/home.htm. The only category not relevant to this comparison is private helicopter and airline pilots.

8. A STAR TO STEER HER BY

1. Homer, *The Odyssey*, trans. Emily Wilson (New York: W.W. Norton), 2018, 5, 270–77.

2. Two constellations Homer mentions—Pleiades and Orion—appear in an older Mesopotamian star catalog, but the other two do not and may have been

Minoan Greek in origin. John H. Rogers, "Origins of the Ancient Constellations: I. The Mesopotamian Traditions," *Journal of the British Astronomical Association* 108, no. 1 (1998): 9–28; John H. Rogers, "Origins of the Ancient Constellations: II. The Mediterranean Tradition," *Journal of the British Astronomical Association* 108, no. 2 (1998): 79–89.

3. Kochab is Beta Ursae Minoris. Polaris is Alpha Ursae Minoris. For China, see Xiaochun Sun and Jacob Kistemaker, *The Chinese Sky During the Han: Constellating Stars and Society* (Leiden, The Netherlands: Brill, 1997). Because the Phoenicians traveled in a narrow north-south zone, they likely opted for the single seemingly stationary star to provide their sailors with an approximation of their position north or south. At other times, no star lay close to the north celestial pole. Pytheas reported an absent pole star in his northern journey in 320 BCE, and classical Roman writings make few references to such a light. "Concerning the north pole Eudoxos is wrong when he says this: there is a certain star that always remains in the same place, and this star is the pole of the world. There is no one star at the pole; it is an empty place, around which three stars turn. The boundary of these encloses a shape at the pole almost like a tetragon, as Pytheas of Marseille also says." Lionel Scott, *Pytheas of Massalia Texts, Translation, and Commentary* (New York: Routledge, 2022), 40. For Hipparchus's *Commentary on the Phainomena of Aratos and Eudoxos*, see Barry W. Cunliffe, *The Extraordinary Voyage of Pytheas the Greek* (London: Allen Lane, 2001), 63–64, 100.

4. Albuquerque, *Curso de História da Náutica* (Coimbra, Portugal: Livraria Almedina, 1972), 45–91.

5. Antoniotto Usodimare, "Carta de Antoniotto Usodimare," in *Viagens dos descobrimentos*, ed. José M. Garcia, 139–44 (Lisbon: Editorial Presença, 1983), 139–44. *Tramontana* is the traditional Italian name for the north. The Portuguese editor, Garcia, identifies the name (albeit without explanation) on p. 144, n9.

6. "Non hauessemo piu che vna volta visto dela tramontana questo perche la ne parea molto bassa sopra del mar e pero la conuegnieuemo uder per tempo molto chiaro e quando la vedessemo ne parea sopra el mar lalteza de vna lanza." Cadamosto, 58.

7. Usodimare, "Carta de Antoniotto Usodimare," 139–44; Cadamosto, 58.

8. "Par la tramontana lalteza de vm homo sopra el mar" (around five feet). Cadamosto, 77.

9. D.A. Shcheglov, "Eratosthenes' Parallel of Rhodes and the History of the System of Climata," *Klio* 88, no. 2 (2006): 351–59; Otto Neugebauer, *A History of Ancient Mathematical Astronomy* (Berlin: Springer, 1975), 725–31; D.A. Shcheglov, "Ptolemy's System of Seven Climata and Eratosthenes' Geography," *Geographia Antiqua* 13 (2004): 21–37.

10. Many excellent works tell how to use Polaris to determine the location and recount the subsequent switch to using the sun for the same purpose. Unfortunately, few draw connections to the mapping of the area, except for the dispute over the method for determining the equator. A partial list includes the many excellent works by Luís de Albuquerque—such as *A naútica e a ciência em Portugal: notas sobre as navegações* (Lisbon: Gradiva, 1989); *Para a história*

da ciência em Portugal (Lisbon: Livros Horizonte, 1973); and *Historia de la navegación portuguesa* (Madrid: Editorial MAPFRE, 1991)—as well as W.L. Randles, "The Emergence of Nautical Astronomy in Portugal in the XV Century," *Journal of Navigation* 51 (1998): 46–57.

11. G.V. Raghava Rau, *Scripture of the Heavens: The Eye of the Heavens* (West Bengal, India: Ananda Press, 1949), 7–8. David King identifies four: sine, universal, horary, and astrolabe/almucantar quadrant. David A. King, *Islamic Astronomical Instruments* (London: Variorum Reprints, 1987).

12. "Ego habebam quadrantem, quando iui ad partes istas, et scripsi in tabula quadrants altitudinem poli arctici, et ipsum meliorem inueni quam cartam." *CVF*, 284. "Quod vellem esse prims qui intraret terram, et sic feci et nullam notitiam habrimus ibi de aliquo homine." *CVF*, 284.

13. *CVF*, 200; U.S. Hydrographic Office, *Sailing Directions for the West Coasts of Spain, Portugal, and Northwest* (Washington, DC, 1908), 119.

14. *CVF*, 200.

15. Andrés de Li and Valentim Fernandes, *Reportório dos tempos. . . .* (Geneva: n.p., 1919), cited by Luís de Albuquerque, ed., *Guía Náutica de Muníque e Guía Náutica de Évora* (Coimbra, Portugal: Imprensa de Coimbra, 1991), 25–26. Relying on the surviving written records alone, Albuquerque dates the origin of this technique to 1450 or 1455 at the earliest. Albuquerque, *Historia de la navegación*, 31. Dispute over leagues per degree (16⅔ or 17½): Albuquerque, *Guía Náutica de Muníque e Guía Náutica de Évora*, 130.

16. "Hauessemo vista de y stele basse sopra el mar chiare e lucida e grande. E tolte queste a segno del busolo ne stauano drieto per ostro afigurado in questo modo seguente . . . le qual I zudigassemo quelle esser el caro del ostro." Cadamosto, 58.

17. "La stella principal nonuedessemo perche el non giera rasoneoule poderla discoprir se prima non perdeuemo la tramontana." Cadamosto, 58–59.

18. Considerable controversy has surrounded the identification of the stars because of the variations in their printed forms. However, the arrangement in the original manuscript shows the most likely combination as alpha and beta Centaurus plus the four main stars of the Southern Cross: α Crucis, β Crucis, γ Crucis, and δ Crucis. Cadamosto, intro. The date was the first week of July, probably the first of the month. The nighttime lasted thirteen hours and the daytime eleven.

19. The *Regimentos do Cruzeiro do Sul* emerged considerably after the sun's use had been established, but it was never very successful. Anon., "Regimentos do Cruzeiro do Sul e de diversas estrelas," in *A marinharia dos descobrimentos*, ed. Abel Fontoura da Costa, 3rd edition (Lisbon: Agência Geral do Ultramar, 1960, 118–47).

20. This remark is as uncharitable to the partisans of longitude as the latter have been to the champions of latitude. The complex history leading to the discovery of a method of using latitude for finding locations at sea appears elsewhere. Longitude is a pair of clocks—period. Latitude requires knowledge of several trigonometric functions and the ability to carry out more than a thousand trig calculations accurately to four decimal places. It also requires a scientifically accurate new instrument and precise measurements at an exact time of day, the

correct day of the year, plus the ability to put the measurements and calculations together. Versus two clocks. See also https://www.ruf.rice.edu/~feegi.

21. José Chabás and Bernard Cohen, "Abraham Zacut and the Transition from Manuscript to Print, Astronomy in the Iberian Peninsula," *Transactions of the American Philosophical Society* 19, no. 2 (2000): 95–98.

22. The longitude measure known as the lunar distance model continued to be used by the British navy well into the early nineteenth century. Captain James Cook used this method during his first journey to the Pacific in 1769, trialing the outrageously expensive mechanical clocks only on his second journey. Nicholas Thomas, *Cook: The Extraordinary Voyages of Captain James Cook* (London: Walker, 2003), 164.

9. THE DEEPEST RIVER AND THE OLDEST DESERT

1. Barros says that Cape Catherine was the last place discovered in the lifetime of King Afonso. The first pillar was put on the south bank of the Congo River "como quem tomaua posse por parte del rey de toda a costa que leixaua atras" (The column was placed as one does to take possession of all the coast up to that point on the part of the king). The *padrão* was called San Jorge. Barros, *Décadas* 1, bk. 3, chap. 3.

2. Cape Lopez, the westernmost point of southern Africa, lies less than half a degree south of the equator, just north of the Ogooué River's mouth. To the south lie the lagoons Ndogo and Nkomi. "From Cape St Catherine SE and 1/4 E & W 20 leagues to 5 degrees south . . . where there are trees and people, many elephants, and other wild animals." *Esmeraldo* (Évora), fol. 79 and 81v.

3. Winds blow from south and south-southwest at force 3 more than 80 percent of the time, according to each monthly map in U.S. Hydrographic Office, *Atlas of Pilot Charts South Atlantic Ocean*, Publication 105.

4. Four storms in April–May were fifty miles long, with winds up to fifty-eight miles per hour (ninety-three kilometers per hour). Off the coast of Africa, they are referred to as "tornadoes," but since they have nothing in common with the North American weather patterns of the same name, I have chosen the term *tropical storms* to maintain clarity. U.S. Hydrographic Office, *Atlas of Pilot Charts;* Lothar Stramma and Friedrich Schott, "The Mean Flow Field of the Tropical Atlantic Ocean," *Deep Sea Research II* 46, nos. 1–2 (1999): 279–303. *Tropical storm* is also used synonymously by the *Africa Pilot*, vol. 2, chap. 7, sec. 180.

5. The fog and low clouds are documented in Nathalie Philippon et al., "The Light-Deficient Climates of Western Central African Evergreen Forests," *Environmental Research Letters*, no. 3 (2019); *Esmeraldo* (Évora), fol. 81. The people were likely from the Bantu-speaking Myene clans. The freshwater sources were the Nyanga, Conkouati, Kouilou, Ogooué, and Nazareth Rivers.

6. "History of Pointe Noire," https://horizon.documentation.ird.fr/exl-doc/pleins_textes/divers12-05/13243.pdf; *Esmeraldo* (Évora), fol. 81v; Denis Binet et al., "El Niño–like Warm Events in the Eastern Atlantic (6°N, 20°S) and Fish Availability from Congo to Angola (1964–1999)," *Aquatic Living Resources* 14, no. 2 (2001): 99–113.

7. Today's smallest bulk ships hold a minimum of twenty thousand deadweight tons (i.e., carrying capacity). The limit in Cabinda is eight thousand deadweight tons. The small river is the Massabi and the harbor is in present-day Angola. Most of Angola's wealth comes from oil wells off Cabinda.

8. The Congo River's fresh water reached 0.3 meters deep only from April to September, but it more than doubled, to 0.7 meters, the rest of the year. "When the northward extension of the plume varies between 100 and 300 km—and 0.7 m during the austral summer (ONDJ and February–March) when the plume reaches 2°S of latitude (i.e., northward extension of 450 km) . . . only a third of the freshwater input into the domain is carried away by the coastal current. The freshwater thus accumulates within the bulge, which expands (100 km wide). . . . The strong northwestward drift ({{gt}}0.5 m s^{-1}) following the Congo River canyon and (2) an upstream (northward) coastal current (0.2 m s^{-1}) in the vicinity of the Congo River mouth . . . both of these flows turn southwestward and join the recirculating gyre." With characteristic insouciance, Barros declared that they encountered fresh water twenty leagues up the coast. Barros, *Décadas* 1, bk. 3, chap. 3.

9. The "H.B.M.S. Rambler . . . [in] April 1899, found that from about latitude 2° N . . . the color of the sea was a dark olive green, getting browner and more like peat water as the Kongo was approached." U.S. Hydrographic Office, *Sailing Directions*, 1908, 60. From the equator southward, "the tide is semidiurnal. . . . The tidal range is comparatively small, only exceeding 1.8 meters at mean springs in a few river estuaries." *Africa Pilot*, vol. 2, chap. 1, sec. 155.

10. This exceptional northward movement of freshwater outflow reaches as far north as 5°26′S, just south of Cabinda, except during February and March, when it moves westward and can be felt eight hundred kilometers (431 nautical miles) out into the Atlantic. Cléa Denamiel et al., "The Congo River Plume: Impact of the Forcing on the Far-Field and Near-Field Dynamics," *Journal of Geophysical Research: Oceans* 118, no. 2 (2013): 964–89. Arriving at similar but not identical conclusions, using a different methodology, are Yan Kang et al., "Areas of the Global Major River Plumes," *Acta oceanologica sinica* 32 (2013): 79–88; Luiz Bruner de Miranda et al., *Estuarine Physical Oceanography* (Singapore: Springer Nature, 2017). In 1507, Pacheco Pereira wrote simply that "from April to September . . . the current is powerful and can be felt 30 leagues out." By 1908, the U.S. report states that the "Kongo stream . . . imparts a clayey appearance to the water . . . a yellowish olive-green color . . . at the distance of 40 miles it is only partially mingled with that of the sea, while the discoloration caused by the freshwater has been known to extend 300 miles off." See *Africa Pilot*, vol. 2, chap. 1, sec. 152, for a description of the color of the water.

11. Only six of these eight men who traveled as far as the Lelala Falls returned. One of the lost had died of an illness, and the other likely from an accident. One man was probably enslaved, given that only his first name appears: Antão (Anthony), placed second to last among the list of expedition survivors. A cross appears above the names of the last two survivors to die, with the word for "sickness" carved beside the name Gonçalo Álvares. The cataracts of Lelala lie near Matadi, ninety-three miles (150 kilometers) upstream from the Atlantic port of Banana (north side of the mouth of the Congo River). It remains

the farthest point upriver that oceangoing ships can reach. Juergen Runge, "The Congo River, Central Africa," in *Large Rivers: Geomorphology and Management*, ed. Avijit Gupta, 2nd edition (Oxford: Wiley-Blackwell, 2022), 433–56.

12. Cão set up the pillar at -6.067, 12.33.

13. Rui de Pina [1440–1521], *Chronica Deel-rei D. Affonso V* (Lisbon: Escriptorio, 1901–2), vol. 2, chap. 133.

14. "Por memoria d'esta primeira vista no proprio logar em que se primeiro viram está uma coluna de marmore mui alta com o escudo Real de Portugal, que o dito doutor João Fernandez da Silveira, embaixador, que era presente, mandou fazer." Pina, *Chronica Deel-rei D. Affonso V*, vol. 2, chap. 126.

15. Carmen M. Radulet and Rui de Pina, *O cronista Rui de Pina e a "Relação do Reino do Congo": manuscrito inédito do "Códice Riccardiano 1910"* (Lisbon: Comissão Nacional para as Comemorações dos Descobrimentos Portugueses, 1992), 96–97, 134. Nearly identical comments appear in the "Manuscrito inédito do 'Códice Riccardiano 1910'" and in Garcia de Resende [ca. 1470–1536] and Gabriel Pereira, *Chronica de El-rei D. João II*, 3 vols. (Lisbon: Escriptorio, 1902), vol. 1, 150–51.

16. Radulet and de Pina, *O cronista Rui de Pina,* 98–99, 136.

17. Barros, *Décadas* 1, bk. 3, chap. 3, wrote: "E posto que leuáua algũas linguas da gente que tinham descuberta, em nenhũua coisa se poderam entender com esta." Cåo left at the end of September. The journey from the Congo River to Lisbon took at least three months. Given at least two months needed to outfit another boat to return, and not including the duration of the meeting with the king, Cão most likely arrived no sooner than May 1484, and likely somewhat later. Other sources list contradictory lengths. Martin Behaim's globe mentions nineteen months for Cão's journey. Martin Behaim, "Erdapfel 1492," globe, Nuremberg, 1490–92, American Geographical Society Library Digital Collection, https://collections.lib.uwm.edu/digital/collection/agdm/id/1224.

18. Joseph Nye, "Soft Power: The Origins and Political Progress of a Concept," *Palgrave Communication* 3 (2017): 17008; Garcia de Resende, *Chronica de el-rei D. João II*, ed. Gabriel Pereira (Lisbon: Escriptorio, 1902), vol. 2, 155–56; Radulet and de Pina, *O cronista Rui de Pina*, 96–97.

19. Barros gives the name of the rebel community as "Mundéquete," a Portuguese mistransliteration of their rival Bantu Teke people variously spelled as "Anzica" or "Anzicana." Barros, *Décadas* 1, bk. 3, chap. 9. Their location at the Malebo Pool is unmistakable. "Os pouos Mundéquetes que habitam cértas ilhas que estam em hũ grande lago dõde say o ryo Zaire . . . os Mundéquetes . . . éram rebelládos & faziam muyto dano em as térras a eles comarcaãs. . . . Foy causa que se bautizásse el rey nam aquella solemnidáde que elle tinha ordenádo depois que a igreja fosse feita. . . . Partido el rey pera aquella guerra que o apressaua em a qual alguũs dos nossos lá foram . . . Ouve victória." Duarte Pacheco Peireira identifies them as Anzica. *Esmeraldo* (Évora), 82v. See also Jodocus Hondius, "Anzicana . . . populi antropophagi" (map), in *Nova Africae tabula* (Amsterdam: Hondius, 1606); Ebiatsa-Hopiel-Opiele, "Les teke: définition historique des hommes et de leur espace (avant le XVIIIe siècle)," *Muntu: revue scientifique et culturelle du CICIBA* 7 (1987): 32–47; James Green,

"Material Values of the Teke Peoples of West Central Africa (1880–1920)" (PhD diss., University of East Anglia, 2018). The baptism of the ruler of Congo occurred on April 3, 1491.

20. Ivory (elephant tusks): Barros, *Décadas* 1, bk. 3, chap. 3; Mariza de Carvalho Soares, "Por conto e peso": o comércio de marfim no Congo e Loango, séculos XV–XVII," *Anais do Museu Paulista* 25, no. 1 (2017): 59–86. For other formal African vists to Portugal, see Michał Tymowski, *Europeans and Africans: Mutual Discoveries and First Encounters* (Leiden, The Netherlands: Brill, 2020), 162–84.

21. The delta would have extended twenty-five miles from the center of the Congo River to Port Banana on the north side of the river. *Esmeraldo* (Évora), fol. 82v–83. Delta size was measured via ESRI, ArcGISPro. While the outlets to the ocean have since been covered by silt or sand, the collection of islands remains.

22. The Angola Current is discussed later in this chapter.

23. On the continued use of this shell along the coast, see Vitorino Magalhães Godinho, *Os descobrimentos e a economia mundial*, 2nd ed., 3 vols. (Lisbon: Editorial Presença, 1981), vol. 2, 19. Godhinho mistakenly includes the name of the classification system (Lamarck) as part of the dwarf olive's Latin name.

24. Cão may also have taken a speaker of Kimbundu. Both Ambundu and Kimbundu are Bantu languages spoken near the coast.

25. From Padrão (south bank of the Congo River) to Ambrizete/N'Zeto, the northeastern edge of Luanda harbor, is 172.3 nautical miles. *Africa Pilot*, vol. 2, chap. 7, secs. 3, 60, 80, 103; Esmeraldo (Lisbon), fol. 67v–68, states the journey from the Congo River to Luanda as thirty-five leagues and Luanda at nine degrees south—the actual location of Luanda.

26. Lobito lies 210 miles south of Luanda. *Africa Pilot*, vol. 2, chap. 7, secs. 175, 202. The harbor is three miles long and one mile wide, sheltered by a lengthy sand spit. In Portuguese, the original name of Lobito was Ponta das Gamboas. *Esmeraldo* (Évora), fol. 83v.

27. *Esmeraldo* (Évora), fol. 83v. A troublesome haze hung over the beaches, making the coast hard to see. *Africa Pilot*, vol. 2, chap. 7, sec. 140.

28. The hill is called Saint Mary's Bonnet and lies 2.5 miles east of Cape Santa Maria. A small commemorative cross remains atop the "bonnet" to this day. Barros, *Décadas* 1, bk. 3, chap. 3, refers to this pillar by an alternate name, "Saint Augustine," located at an even thirteen degrees. Now removed to Lisbon and replaced by a replica, the pillar's inscription reads: "In 1482, the year of our Lord Jesus Christ, the very excellent, elevated, and powerful prince, King Dom João II ordered the discovery of this land, and placing of these standards by this Diogo Cão, squire of his court." Both base and top crosses are missing, but according to the Geographic Society of Lisbon, the column is five feet five inches (169 cm) high and the rectangular stone column bearing the inscription remains intact.

29. The coastal cliffs between Cape Saint Maria and Cape Saint Marta—consisting of basalt, a dark volcanic rock—disappeared after Cape Saint Marta. Bruce Cairncross, "The Where of Mineral Names: Namibite, Namib Desert, Copper Valley, Khorixas District and Region Namibia," *Rocks & Minerals* 93,

no. 2 (2018): 184–87. In the twenty-eight miles between the two capes lay the Rio Catara, Rio Cangala, and the Baía Santa Marta. *Africa Pilot*, vol. 2, chap. 7, sec. 223. The Carunjamba was originally called Rio São Lorenzo; see *Esmeraldo* (Évora), fol. 83v.

30. "The Angola Current forms the eastern section of a large, cyclonic gyre in the Gulf of Guinea. In the upper layer (0–100 m), it seems to be formed mainly by the southeast branch of the South Equatorial Countercurrent and the southward-turning waters from the north branch of the Benguela Current." P.M. Wedepohl et al., "Surface Drift in the South-East Atlantic Ocean," *South African Journal of Marine Science* 22, no. 1 (2000): 71–79. The current's temporal variability was detected more than forty years ago. During four months of the year, March–April and September–October, at twenty-nine nautical miles (fifty-three kilometers) offshore, the Angola Current's velocity exceeded forty centimeters per second—meanwhile, along the shore, the maximum mean southward speed was eight centimeters per second. See also C.A. Dias, "Preliminary Report on the Physical Oceanography off Southern Angola," *Collection of Scientific Papers International Commission for the Southeast Atlantic Fisheries* 10 (1971): 103–16. In nautical terms, the inshore currents glided at a sluggish 0.15 knots. Kopte and colleagues observed "a southward transport maximum in May, . . . [that] suggests a secondary transport maximum in March/April." R. Kopte et al., "The Angola Current: Flow and Hydrographic Characteristics as Observed at 11°S," *Journal of Geophysical Research: Oceans* 122, no. 2 (2017): 1177–89. U.S. Hydrographic Office, *Sailing Directions for the West Coasts of Spain, Portugal, and Northwest* (Washington, DC, 1908), notes that in the vicinity of Palmeirinhas Point, just south of Luanda, the current might begin to set northward with considerable force, suggesting an unusually northward reach.

31. From Luanda, Angola, to Alexander Bay (south of the Orange River in South Africa), there were land/sea breezes year-round. Matthew Fontaine Maury, "Maury's Wind & Current Chart, [North Atlantic]: No. 2, Series A" (Washington, DC: U.S. Hydrographic Office, 1851), 404; U.S. Hydrographic Office, *Sailing Directions for the West Coasts*, 398–99. Ships sailing further offshore would use the land breezes.

32. Geological research has shown that this region has been the meeting place for these two currents since the late Quaternary. Ralph R. Schneider et al., "Late Quaternary Surface Circulation in the East Equatorial South Atlantic: Evidence from Alkenone Sea Surface Temperatures," *Paleoceanography* 10, no. 2 (1995): 197–219. The "Angola Current flowing south . . . joins the northward-flowing Benguela Coastal Current . . . near 16°S–14°S." Wedepohl et al., "Surface Drift in the South-East Atlantic Ocean," 71–79. See also P.M.F. Colberg and C.J.C. Reason, "A Model Study of the Angola Benguela Frontal Zone: Sensitivity to Atmospheric Forcing," *Geophysical Letters: Oceans* 33, no. 19 (2006); J.M. Meeuwis and J.R.E. Lutjeharms, "Surface Thermal Characteristics of the Angola Benguela Front," *South African Journal of Marine Science* 9, no. 1 (1990): 261–79. Wedepohl et al. write that "alternating periods of southward and northward velocities that can last for a couple of months with superimposed high-frequency velocity pulses" and "no steady southward flow of the AC is observed at ~11°S off Angola. Instead . . . [it features] strong intraseasonal variability." According

to *Esmeraldo* (Évora), fol. 87v, the strong Benguela winds and currents can start as early as April. By August, they shoot almost another two hundred miles farther north of the region where the Angola and Benguela Currents meet.

33. Maury, "Maury's Wind & Current Chart," 402.

34. Between fifteen and twenty degrees south and from ten degrees east to the coast, calms occur 10 percent of the time; but 85 percent of the time, winds blow between eleven and sixteen knots from the south and the southeast. U.S. Hydrographic Office, *Atlas of Pilot Charts North Atlantic Ocean (Including the Gulf of Mexico)*, Publication 106 (Washington, DC, 2002), months of June, July, and August.

35. "Atmospheric circulation . . . is controlled by a low-pressure system over Africa moving from the 15°W meridian toward equatorial regions during the austral summer (October to March) and toward the Sahara during the austral winter (April to September). The 16°S parallel is approximately the northern boundary of the zonally directed trade wind field . . . influenced by the low-pressure system." Ralph R. Schneider et al., "Late Quaternary Surface Circulation in the East Equatorial South Atlantic: Evidence from Alkenone Sea Surface Temperatures," *Paleoceanography* 10, no. 2 (1995): 197–219. The Angola-Benguela front intersection is near sixteen degrees south. Wedepohl et al., "Surface Drift in the South-East Atlantic Ocean," 71–79. A more precise location describes an anticyclonic South Atlantic "between 45°S–15°S and 55°W–10°E . . . forced by wind-driven Ekman convergence." Kimberley L. Drouin et al., "Variability and Trends of the South Atlantic Subtropical Gyre," *Journal of Geophysical Research: Oceans* 126 (2021): e2020JC016405. Subtropical high pressure in the area was noted in Maury's mid-nineteenth-century compilation. Maury, "Maury's Wind & Current Chart," 402.

36. The first seventy-two miles south-southwest after the Carunjamba River brought two seasonal rivers that supplied fresh water: the Inamagando and Bentiaba (São Nicolau). Water may also have been available at the Baía das Salinas, Baía Velha, and Baía do Baba. *Africa Pilot*, vol. 2, chap. 7, secs. 232, 236, 239. Some sources define the desert as beginning at Namibe Bay, seventy-plus miles to the south, rather than the Carunjamba River. Given the immense length of the desert, the difference between these two points is relatively small.

37. This location, with its ample fishing grounds, may have been another area from which Cão took one or more individuals as translators.

38. Duarte Pacheco Peireira's latitude for Namib (Moçâmedes) Bay is surprisingly accurate: It is fifteen degrees, twenty minutes (−15.3333), off by about eight miles (actual latitude: −15.195278).

39. "A terra derredor he de muita area e elle tem hum mato baixo rraso," *Esmeraldo* (Évora), fol. 84; *Africa Pilot*, vol. 2, chap. 7, sec. 262. Haversine distance to Cabo Negro thirty-two nautical miles. Rio dos Flamingos lay ten nautical miles north of Cabo Negro. The pillar was removed to the Sociedade de Geografia (Lisbon) in 1892, and a replica was placed. From Pointe-Noire to Moçâmedes (Namib), the heavy surf made anchoring in shallow bays dangerous. Maury, "Maury's Wind & Current Chart," 403. Even the Namibe harbor was open to occasionally heavy swells from the southwest and winds from the

north and northeast. Fortunately, tides were virtually nonexistent. *Africa Pilot*, vol. 2, chap. 7, secs. 246, 251–52.

40. *Africa Pilot*, vol. 2, chap. 7, sec. 269. Namibe to Kunene 132 miles: *Africa Pilot*, vol. 2, chap. 7, secs. 258, 265.

41. *Esmeraldo* (Évora), fol. 84v, describes the roofs as made of sand, but no one investigated the roofing material. Whale bones often appeared in Inuit homes, but there are few reports from southwestern Africa. Andrew B. Smith and John Kinahan, "The Invisible Whale," *Coastal Archaeology* 16, no. 1 (1984): 89–97.

42. The inhabitants of Tombua harbor had to transport water from the north (Rio Curoca) or the south (Rio Kunene). *Esmeraldo* (Évora), fol. 84v, says that the coast by Baía dos Tigres is difficult to sail from April to September.

43. *Africa Pilot*, vol. 2, chap. 7, secs. 268, 269.

44. The Moçâmedes Desert, fifty-five to eighty million years old, extends south along the Atlantic coast of Angola to the Angola-Namibia border for 275 miles (450 kilometers). The Kaokoveld Desert occupies a coastal strip covering 45,700 square kilometers (17,600 square miles), from thirteen to twenty-one degrees south. The Atlantic Ocean bounds it to the west. The British Hydrographic Office and the U.S. military locate the Skeleton Coast between the Rio Kunene (the southern boundary of Angola) and Walvis Bay, adding that the name comes from the whale skeletons. *Africa Pilot*, vol. 2, chap. 8, sec. 5; U.S. Hydrographic Office, *Enroute (SW Coast of Africa)* (Washington, DC, 2021), 147. Barros, *Décadas* 1, bk. 3, chap. 3, dismissed it as a "manga das areas."

45. The closest evidence of any human settlement occurred at a single isolated hill seventy-five miles (120 kilometers) inland southeast of the mouth of the Curoca River. The Tchitundu-Hulu mountain showed signs of sporadic human settlement between the first millennia BCE and CE, when several seasonal rivers appeared along the hill. Humans seem to have abandoned the region when the water vanished. A.R. Willcox, *The Rock Art of Africa* (London: Routledge, 2016), 46–47.

46. Andrew Goudie, "Namib Sand Sea: Large Dunes in an Ancient Desert" *Geomorphological Landscapes of the World*, ed. Piotr Migoń (New York: Springer, 2010), 163–69. For recent changes, see Andrew S. Goudie and Heather A. Viles, *Geomorphology in the Anthropocene* (London: Cambridge University Press, 2017); Brigitte Senut and Loïc Ségalen, "Neogene Palaeoenvironments of the Namib Desert: A Brief Synthesis," *Transactions of the Royal Society of South Africa* 69, no. 3 (2014): 205–21. The gymnosperm *Welwitschia mirabilis* is the sole surviving member of its family. Dieter J. von Willert et al., *Life Strategies of Succulents in Deserts: With Special Reference to the Namib Desert* (Cambridge: Cambridge University Press, 1992), 77. Prevailing local southwesterly winds, cooled by the Benguela Current along the west coast of South Africa, produce an inland movement of cold air, which is overlain by the warmer, lighter air. This inversion prevents the convectional rise of the cool, humid air, with the result that no clouds form. The cool air does, however, regularly form a stable layer of fog, which is blown inland as far as thirty-seven miles, or sixty kilometers. H. Andersen et al., "Spatiotemporal Dynamics of Fog and Low

Clouds in the Namib Unveiled with Ground- and Space-Based Observations," *Atmospheric Chemistry and Physics* 19 (2019): 4383–92.

47. Maury, "Maury's Wind & Current Chart," 404. From southern Angola south, occasional hot winds from inland east or northeast can reach twenty knots and produce the sensation of sandpaper. Usually they last a few days, but they can continue for many more. Although most common during spring and fall, these winds can occur anytime.

48. Dieter J. von Willert et al., *Life Strategies of Succulents in Deserts: With Special Reference to the Namib Desert* (Cambridge: Cambridge University Press, 1992).

49. The land/sea breezes continued in Alexander Bay year-round. Maury, "Maury's Wind & Current Chart," 404. The safe distance to sail is ten miles from the shore. *Africa Pilot*, vol. 2, chap. 8, sec. 5. "The minimum speeds for the southern regions of the [Benguela] current are found during winter (10 $cm{\cdot}s^{-1}$) and the maximum during summer (18 $cm{\cdot}s^{-1}$). In the central regions, the speeds are minimum during winter (9 $cm{\cdot}s^{-1}$) and maximum during summer (16 $cm{\cdot}s^{-1}$). In the northern region, speeds are maximum during winter (10 $cm{\cdot}s^{-1}$) and minimum during summer (7 $cm{\cdot}s^{-1}$)." Wedepohl et al., "Surface Drift in the South-East Atlantic Ocean," 78.

50. The flat sandy area represented the four major "ephemeral rivers" that had disappeared deep into desert sands hundreds of kilometers inland. The first transient river included the Houarusib (128 nautical miles beyond the Kunene), followed by the Uniab (204 nautical miles), Huab (243 nautical miles), and Ugab (268.7 nautical miles).

51. A fifth such fleeting river, the Messum, 275 nautical miles south of the Kunene, had freshwater pools near its mouth, but it is unclear whether that would have proved sufficient for the vessels. Cão's first possible anchorage to search for water lay nearly 130 nautical miles along the Skeleton Coast at Angra Rui Pires (now Mowe Bay) or slightly farther at São Amaro (now Palgrave Point). From Rio Kunene to Cape Cross is an estimated 308 nautical miles. *Africa Pilot*, vol. 2, chap. 8, secs. 16–27.

52. Gales, active 10 percent of the time, begin offshore at twenty-one degrees south and eight degrees east but start within two degrees of the coast, at 120 miles, in January; 10 percent rough at twenty-two degrees south in January. U.S. Hydrographic Office, *Atlas of Pilot Charts*.

53. Barros erroneously has Cão ending his journey at Angra do Salto, where he captured two individuals to return to Portugal. Barros, *Décadas* 1, bk. 3, chap. 4. Even if he had captured fishermen from the last human settlement he encountered (Tombua), that place lay considerably north of the pillar Cão left inscribed with his name.

54. Originally it was known by the same name as the pillar on the south side of the Congo River.

55. *Esmeraldo* (Évora), fol. 84v–85r. Barros says the pillar (padrão) inscriptions are in Portuguese and Latin, but only Cape Cross has these two languages. *Esmeraldo* (Lisbon), 66v, lists the more common trio of languages: Portuguese, Latin, and Arabic.

56. The best analysis of the controversy over the exact place where Cão died appears in Eric Axelson, *Portuguese in South-East Africa 1488–1600* (Cape Town: Struik, 1973), 10 n57.

57. Barros, *Décadas* 1, bk. 3, chap. 3, characterizes the distance Cão traveled as two hundred leagues beyond the Congo River and mentions only two pillars: one called "Saint Augustine" at thirteen degrees south, and a second called "Pillar Point" at twenty-two degrees south. As Barros puts it, "Junto da manga das areas." He omitted Cabo Negro. Barros, *Décadas* 1, bk. 3, chap. 4.

58. Vasco Fernandes de Lucena and Francis M. Rogers, *The Obedience of a King of Portugal*, trans. and ed. Francis M. Rogers (Minneapolis: University of Minnesota Press, 1958), 47–48.

59. Carmen Radulet, "As Viagens de descobrimento de Diogo Cão: Nova proposta de interpretação," *Maré Liberum* 1 (1990): 195.

60. Vasco da Gama came from a noble family and was hence granted the additional honorific *Dom*, which in Portugal, unlike in Spain, was used only for members of the royal family.

61. "[N]os praz de nosso propreo moto ciencia poder absoluto e separamos como de feito o separamos do número de plebeo e o habilitamose fazemos nobre de cota de armas." *PMA*, 318. Only three other men connected to the journey to Africa in the fifteenth century received coats of arms: the pilot Nicolau Coelho, the director of operations on the Guinea coast Fernão Gomez, and the head of the Mina fort's construction, Diogo de Azambuja. Da Gama received a title in 1519. *Livro do Armeiro-Mor* (Lisbon, n.p., 1509).

62. António José Saraiva, *O crepúsculo da Idade Média em Portugal* (Lisbon: Gradiva, 1990), 253.

10. A FIRST GLIMPSE OF THE INDIAN OCEAN

1. Dias's effort was not the first to venture farther south. A previous attempt had been forced to return because of the lack of available fresh water. Thus, Dias's expedition was given a water supply vessel, the first time such a step had become necessary. J. Semedo de Matos, "A caravela de Nuño Tristão: A caravela de descobrir," *Revisita de Armada*, no. 218 (1997), 21.

2. According to João de Barros, Cão discovered 355 leagues of coast, while Dias discovered 350. These numbers reflect the degree of recognition accorded each of the two leaders, not actual figures. Barros, *Décadas* 1, bk. 3, chaps. 4 and 5.

3. "Alem de assentãrem os padrões que leuáuãnas distâncias do comprimento da costa que ilhe bem parecia, eram postos em lugáres notáués." Barros, *Décadas* 1, bk. 3, chap. 4.

4. Today, this dune extends 11.5 miles, enclosing a large lagoon.

5. Barros, whose latitudes are generally fairly accurate, places Serra Parda at twenty-four degrees south. Modern-day calculations place Conception Bay at 23.91 south. *Africa Pilot*, vol. 2, chap. 8, secs. 3, 74. Axelson identifies the Golfo de Conceção on the Cantino map as Walvis Bay. However, a place labeled "Conception Bay" appears on the Cantino map, along with a *padrão* design.

6. *CVF*, fol. 87.

7. The replacement pillar appears in the exact location where fragments of the original pillar were located.

8. C.L. Griffiths et al., "Impacts of Human Activity on Marine Animal Life in the Benguela: A Historical Overview," in *Oceanography and Marine Biology: An Annual Review*, ed. R.N. Gibson et al. (Boca Raton, FL: CRC Press, 2004), vol. 42, 303–92; R. Simmons, "The Status of Coastal Wetlands in Namibia," in *Wetlands Conservation for Southern Africa*, ed. T. Matiza and H.N. Chabwela (Gland, Switzerland: International Conservation Union, 1992), 125–52; A. Hjort et al., *Climate Change Implications for Fisheries of the Benguela Current Region: Making the Best of Change* (Rome: Food and Agricultural Organization of the United Nations, 2012), 87; R.J.M. Crawford et al., "The Benguela Ecosystem IV: The Major Fish and Invertebrate Resources," *Oceanography and Marine Biology* 25 (1987), 305.

9. Large pillar fragments were recovered in the nineteenth century, and a replica was made to replace the original. Despite being one of few large anchorages along the coast, Lüderitz lacked links to trade routes and never became a significant port until the end of the nineteenth century.

10. The area covered the coast from Bakersbay or Lion's Head Cape fur seal colony. Such colonies only appeared in areas with plentiful fish.

11. "Punham tambem os nomes aos cábos angras & móstras da terra que descobriam, ou por razam do dia que aly chagáuam, ou por qualquer outra causa." Barros, *Décadas* 1, bk. 3, chap. 86. The Cantino map of 1502 provides a great many names, none of which can be absolutely identified today. Anonymous, Cantino chart (Modena, Italy: Biblioteca Estense, 1502).

12. Barros, *Décadas* 1, bk. 3, chap. 86, gives the location as twenty-nine degrees south (actually Holgat). Eric Axelson locates the departure point farther south, near the Cedarburg Mountains (formerly Serra dos Reis). Eric Axelson, *Portuguese in South-East Africa 1488–1600* (Cape Town: Struik 1973), 14. However, the point with a good supply of clear water would have been in the harbor at Holgat.

13. Around fifteen to twenty degrees south, five to ten degrees east, winds remain calm only one percent of the time and at force 3 fifty percent of the time. Hence, sailing within 580 miles of the coast would have been difficult, and Dias may have sailed farther out. However, in January and slightly farther south (twenty-two degrees), the rough winds decreased to only 10 percent of the time. Barros, *Décadas* 1, bk. 3.

14. The current is the South Atlantic Current, which flows east between about forty-three and forty-seven degrees south, north of the subantarctic fronts. See A. Cabré et al., "Subtropical-Tropical Transfer in the South Atlantic Ocean," *Journal of Geophysical Research: Oceans* 124 (2019): 4820–37. Meteorological data from U.S. Hydrographic Office, *Enroute (SW Coast of Africa)*. The pilots were Pero d'Alenquer, João de Santiago, and Alvaro Martinez.

15. "Despite its importance, in-situ observations are sporadic and sparse in time and space in the S[outh] A[tlantic]." Xue Wang et al., "Characteristics of 3-Dimensional Structure and Heat Budget of Mesoscale Eddies in the South Atlantic Ocean," *Journal of Geophysical Research: Oceans* 126, no. 5 (2021). Furthermore, anticyclonic eddies and cyclonic eddies "are frequently observed

south of ~30°S for the studied period, where the eddy frequency reaches from 20% to 50%. In contrast, eddy frequency drastically weakens in the subtropical S[outh] A[Atlantic] between 10°S and 20°S and is less than 5%."

16. This is only an estimate, since the range of possible longitudes at thirty-five degrees south is too large. For November–January, see U.S. Hydrographic Office, *Pilot Chart of the South Atlantic*, Publication 105, 2009.

17. For Beaufort wind scales, see U.S. Hydrographic Office, *Pilot Chart of the South Atlantic*. This suggests that his journey into the Atlantic may have extended as far as somewhere between those two numbers, but a final judgment is impossible.

18. Cape Agulhas lies ten minutes of arc south of Dias's reported position. West winds of force 4 were strongest between ten and fifteen degrees east at thirty-five degrees south. At that latitude, he would have been between 8.43 and 3.43 degrees or between 205.8 and 505.8 nautical miles west of Cape Town. With currents between seventeen and twenty-one knots, and traveling due west, he might have been as few as ten hours away from catching sight of the coast.

19. At this point, accounts differ. Some argue that he found it impossible to anchor nearby and that the following episode occurred in Mossel Bay. Velho, 9–10; Barros, *Décadas* 1, bk. 3, chap. 4. There are three well-known transcriptions: Alvaro Velho. *Roteiro da primeira viagem de Vasco Da Gama*, ed. Neves Águas (Mem Martins, Portugal: Publicacões Europa-América, n.d.); Alvaro Velho and Abel Fontoura da Costa, *Roteiro da primeira viagem de Vasco da Gama (1497–1499)*, with preface and notes by A. Fontoura da Costa, 3rd edition (Lisbon: Agência-Geral do Ultramar, 1969); Alvaro Velho, *Roteiro da primeira viagem de Vasco de Gama à Índia, 1497–1499 = Journal of the First Voyage of Vasco da Gama to India, 1497–1499*, ed. Luís Fernando de Sá Fardilha and Maria de Lurdes Correia Fernandes (Porto, Portugal: U. Porto Edições, 2016). Although controversy surrounds the actual authorship, I have opted for the traditionally used Alvaro Velho and the version listed first.

20. Velho, 10. The maximum distance for an effective medieval crossbow to kill a target is 320 meters.

21. I follow Eric Axelson's locations here, since he grew up riding with his captain father on a boat hauling freight between different places along the South African coast. Personal communication, June 1997. Other Portuguese names included Serra da Estrela, Punta de Pescaria, Bahia das Alagoas, Ponta das Qeuimadas, and Cabo da Roca.

22. Barros, *Décadas* 1, bk. 3, chap. 87; Gabriel Pereira [1847–1911], ed., *Roteiros Portuguezes Da Viagem De Lisboa à India Nos Seculos XVI E XVII* (Lisbon: Imprensa nacional, 1898).

23. Barros, *Décadas* 1, bk. 3, chap. 86.

24. Barros, *Décadas* 1, bk. 3, chap. 4. For da Gama's battle with the Agulhas, see chapter 11.

25. The legend appears in Barros, *Décadas* 1, bk. 3, chap. 4. The *Esmeraldo* does not mention the legend. *Esmeraldo* (Évora), 111.

26. Barros, *Décadas* 1, bk. 2, chap. 4, wrote that Dias placed "o quinto [padrão] Sam Felipe, no grande & notauel cábo de boa esperança." Eric

Axelson, who spent decades looking for traces of the Portuguese limestone, never found remnants on the cape.

27. From Cape Town to Wallis Bay, a land breeze around eight to twelve knots prevailed from June to September, and a sea breeze from December to March. However, about halfway between Cape Town and Walvis Bay, the sea breeze sometimes reached gale-force winds between October and March. The halfway point was Alexander Bay, just south of the Orange River.

28. Barros, *Décadas* 1, bk. 3, chap. 4.

29. "E a razam que déram dos mortos foy fiarense dos negros da térra com quem viéram tér commincaçam: os quaes sóbre cóbiça da algũas cousas q~ resgatáuam os matáram." Barros, *Décadas* 1, bk. 3, chap. 88.

30. Barros, *Décadas* 1, bk. 3, chap. 88.

31. Barros, *Décadas* 1, bk. 3, chap. 85; C.F. Beckingham, "The Quest for Prester John," in *The European Opportunity*, ed. Felipe Fernandez-Armesto (London: Routledge, 1995).

32. "Porque como éram molhéres com quẽ os hómeẽs nã tem guerra, nam ilhes auiamde fazer mal algum." Barros, *Décadas* 1, bk. 3, chaps. 85–86.

33. "Por nam virem com as mãos vazias." Barros, *Décadas* 1, bk. 3, chap. 88.

34. For Dias's version, see *Esmeraldo* (Évora), fol. 87–83; *Esmeraldo* (Lisbon), fol. 74–77. See also Manuel de Mesquita Perestrelo, *Roteiro da África do sul e sueste desde o Cabo da Boa Esperança até ao das Correntes, 1576*, ed. A. Fontoura da Costa, trans. George McCall Theal (Lisbon: Agência Geral do Ultramar, 1939), 24.

35. Francisco Contente Domingues, *Os navios do mar oceano: teoria e empiria na arquitectura naval portuguesa dos séculos XVI e XVII* (Lisbon: Centro de História da Universidade de Lisboa, 2004), 243–52.

36. Da Gama's initial fleet included a boat captained by Bartolomeu Dias but headed along the familiar route to the gold-and-slave fort at Mina, rather than taking the trans-Atlantic passage. Velho, 2.

11. CRISSCROSSING THE ATLANTIC

1. See chapter 4. For why this same issue did not confront Polynesian sailors whose only voyage north of the equator led to Hawaii, see the introduction.

2. *Corpo Chronológico, maço* 1, parts 1–3, in Armando Corttesão, ed., *O Mistério de Vasco da Gama* (Coimbra, Portugal: Junta de Investigacoes do Ultramar, 1973).

3. The reasons for the treaty are usually attributed to Columbus, whose only discovery at the time was a small set of Caribbean islands. Another decade would pass before any Spanish ship discovered the land to the west.

4. The Évora (99v) and Lisbon (79v) copies of *Esmeraldo* both falsely claim that a ship need only sail straight south from the Cape Verde Islands for six hundred leagues before reaching nineteen degrees south. While correctly demonstrating knowledge of the opposite directions in the two hemispheres, Pereira oversimplified the task of crossing equatorial waters. However, both *Esmeraldo* manuscripts have the correct direction of travel onward from nineteen degrees: east-southeast.

5. The North Equatorial Current moved between twenty and five degrees north. The direction reversal (to a westward flow) appeared between eighteen and twenty-two degrees west.

6. The North Atlantic Equatorial Countercurrent traveled between five and eight degrees north. Miquel Rosell-Fieschi et al., "Zonal Jets in the Equatorial Atlantic Ocean," *Progress in Oceanography* 130 (2015): 1–18; Philip L. Richardson and David Walsh, "Mapping Climatological Seasonal Variations of Surface Currents in the Tropical Atlantic Using Ship Drifts," *Journal of Geophysical Research: Oceans* 91, no. C9 (1986): 10537–50.

7. The northernmost "southern" current flowed slowly westward north of the equator between three and two degrees north.

8. Bernard Barnier et al., "On the Seasonal Variability and Eddies in the North Brazil Current: Insights from Model Intercomparison Experiments," *Progress in Oceanography* 48, nos. 2–3 (2001), 195–230. The current's volume is 36 sverdrup (Sv) north of the equator and 21 Sv from five to ten degrees south. Ilson C.A. da Silveira et al., "On the Origins of the North Brazil Current," *Journal of Geophysical Research: Oceans* 99, no. C11 (1994): 22501–12.

9. Martha W. Buckley and John Marshall, "Observations, Inferences, and Mechanisms of the Atlantic Meridional Overturning Circulation: A Review," *Reviews of Geophysics* 54, no. 1 (2016): 5–63.

10. Given the lack of separation between the Brazil and Caribbean currents, Portuguese ships could have easily wound up in the Caribbean by mistake.

11. "The surface SSEC bifurcation occurs around 14°S in annual mean." C.D. Luko et al., "Revisiting the Atlantic South Equatorial Current," *Journal of Geophysical Research: Oceans* 126,. no. 7, (2021): e2021JC017387.

12. Named for a nineteenth-century Frenchman, Gaspard-Gustave Coriolis. Anders Persson, "How Do We Understand the Coriolis Force?," *Bulletin of the American Meteorological Society* 79, no. 7 (1998): 1373–86.

13. In descending order: Royal-Charlotte Bank, Vitória-Trinidade Ridge, and Abrolhos Bank. C.D. Luko et al., "Revisiting the Atlantic South Equatorial Current"; F. Negrão et al., "The First Biological Survey of the Royal Charlotte Bank (SW Atlantic) Reveals a Large and Diverse Ecosystem Complex," *Estuarine, Coastal and Shelf Science* 255 (2021): 107363.

14. "XXII do dito mês hindo na volta do mar ao sull e a quarta do sudueste, achámos muitas aves feitas como garçõees, e quando vêo a noute tiravam contra o susoeste muito rrigas como aves que hiam pera terra, e neste mesmo dia vimos huuma baléa." Velho, 3. Male giant frigate birds have a green sheen in the sunlight; females have a purple sheen. Bruno de Andrade Linhares, "Aves Marinhas Como Transportadoras de Matéria Para o Ecossistema Insular do Arquipélago dos Abrolho" (PhD diss., Institute of Oceanography, Universidade Federal do Rio Grande, 2022), 16.

15. Negrão et al., "The First Biological Survey of the Royal Charlotte Bank." But no whales—the next area was an uplift from the seafloor and hence not rich in marine animals.

16. José Manuel Malhão Pereira has Vicente Rodrigues's 1577 *Roteiro* as the first of many to mention Abrolhos. However, da Gama's vessels described the area in detail eighty years before.

17. Triggerfish in Abrolhos Reef.

18. Thirty-nine to twenty-seven degrees west. Negrão et al., "The First Biological Survey of the Royal Charlotte Bank."

19. Velho, 8. Pero d'Alanquer, a pilot who had previously sailed around the cape, hugely underestimated their distance. They were still far from the South African coast; in reality, they were thirteen degrees north and forty-seven degrees west.

20. The estimate of the remaining distance uses figures provided in *Esmeraldo* (Lisbon), 511.

21. U.S. Hydrographic Office, *Sailing Directions Enroute*, 2019.

22. Velho, 3.

23. Velho, 3–4.

24. Velho, 4.

25. Velho, 5. The actual first contact took persuading a diminutive man who was collecting honey from the many beehives to eat with them, before returning him to land. Identification from archeological evidence reported in Andrew B. Smith, "Exploitation of Marine Mammals by Prehistoric Cape Herders," *South African Journal of Science* 89 (1993): 161–65. Velho confirms the eating of a cooked seal.

26. Abalone shells "shine like silver." Velho, 6. Tiago Gil Curado, "O Ceitil. Um símbolo da Modernidade portuguesa e a sua evolução estética (1440–1570)," *Documenta & Instrumenta* 17 (2019): 19–31.

27. An *azagaya* is a short wood-throwing spear with an iron tip (javelins are designed for long-distance use). Velho, 6.

28. "Em este ilhéu há muitos lobos marinhos e [alguns] deles são tão grandes como ursos muito grandes, e são muito e têm muito grandes dentes; e vêm aos homens e nenhuma lança, por [mais] força que leve, os não pode ferir. E outros [são] mais pequenos e outros muito pequeninos; e os grandes dão urros como leões e os pequeninos como cabritos." Velho, 14. Off Seal Island, they were still 955 nautical miles (1,769 kilometers or 1,099 miles) from Port Elizabeth, which lies on the other side of the South African coast.

29. "E aqui fomos um dia a folgar e vimos, entre grandes e pequenos, obra de três mil; e atirávamos-lhes do mar com bombardas. E neste ilhéu há umas aves que são tamanhas como patos, e não voam porque não têm penas nas asas, e chamam-lhes soliticairos e matámos delas quantas quisemos; as quais aves zurram como asnos." Velho, 14.

12. ENCOUNTERS ALONG THE AFRICAN COAST AND IN INDIA

1. Gabriel Ferrand, "Le pilote arabe de Vasco de Gama et les instructions nautiques des arabes au XVe siècle," *Annales de géographie* 31, no. 172 (1922): 289–307.

2. The journal of the da Gama fleet's trip to India has been reprinted multiple times. The currently most popular version embroiders the account from the day da Gama acquired his first pair of Indian Ocean pilots, including such inappropriate details as his presumed emotions and excessive detail about interactions on land. Since such an account lacks the characteristics of legitimate Por-

tuguese logbooks, I have sidelined this version in favor of the more skeletal, ocean-oriented account. Furthermore, the embellishments attached to the text cover only the period when the fleet entered seas that had been sailed by thousands of ships for hundreds of years. Hence, sailing this area required no navigational innovations, making those sections of no consequence to this book. I have used *Roteiro da primeira viagem de Vasco Da Gama*, ed. Neves Águas (Mem Martins, Portugal: Publicacões Europa-América, n.d.). Other editions include Alvaro Velho and Abel Fontoura da Costa, *Roteiro da primeira viagem de Vasco da Gama (1497–1499)*, with preface and notes by A. Fontoura da Costa, 3rd edition (Lisbon: Agência-Geral do Ultramar, 1969); Alvaro Velho, *Roteiro da primeira viagem de Vasco de Gama à Índia, 1497–1499 = Journal of the first voyage of Vasco da Gama to India, 1497–1499*, ed. Luís Fernando de Sá Fardilha and Maria de Lurdes Correia Fernandes (Porto, Portugal: U. Porto Edições, 2016). See Velho, 9–10.

3. Velho, 11.

4. Velho, 12.

5. Velho, 12–13.

6. Velho, 14.

7. From Afrikaans *berg* (mountain).

8. "Da espessa nuvem setas e pedradas / Chovem sobre nós outros sem medida"; canto 5, stanza 33 / "Bramindo o negro mar, de longe brada / Como se desse em vão nalgum rochedo"; canto 5, stanza 38. Luís de Camões [1524?–80], *Os Lusiadas: Poema Epico* (Lisbon: Parceria Antonio Maria Pereira, 1928). Compare these lines of Wordsworth: "I dropped my pen; and listened to the Wind / That sang of trees uptorn and vessels tost."

9. Michael A. Barnes and Christo Rautenbach, "Toward Operational Wave-Current Interactions over the Agulhas Current System," *Journal of Geophysical Research: Oceans*, 125, no. 7 (2020): e2020JC016321. The waves must meet at a 120° angle.

10. When wave trains meet at a 120° angle, they break upward. See Tamaryn Morris et al., "The Importance of Monitoring the Greater Agulhas Current and Its Inter-ocean Exchanges Using Large Mooring Arrays," *South African Journal of Science* 113, nos. 7–8 (2017).

11. The number and extent of the sunken ships are unknown, as the naval officer in charge of charting downed vessels acknowledged to me. He added that the wrecks were so voluminous and the scavengers so numerous that no control over the waters and wrecks was possible. Personal communication, July 14, 1998.

12. Dias had named the river for João de Infante, captain of the second boat in his fleet. Pillar identification by Eric Axelson, *Portuguese in South-East Africa 1488–1600* (Cape Town: Struik, 1973).

13. On the wind shifts, see M. Rouault et al., "The Atmospheric Boundary Layer Above the Agulhas Current During Alongcurrent Winds," *Journal of Physical Oceanography* 30, no. 1 (2000): 40–50.

14. From Mossel Bay to Port Elizabeth was 212 nautical miles, from Alagoa Bay to East London was 139 nautical miles, and fron East London to Durban was 255 miles.

15. Even today, smaller boats are warned off from trying to approach the coast. Greta M. Leber et al., "Wind and Current Forcing Combine to Drive Strong Upwelling in the Agulhas Current," *Journal of Physical Oceanography* 47, no. 1 (2017): 123–34.

16. Harry L. Bryden et al., "Structure and Transport of the Agulhas Current and Its Temporal Variability," *Journal of Oceanography* 61, no. 3 (2005): 479–92; I.V. Lavrenov, "The Wave Energy Concentration in the Agulhas Current," *Natural Hazards* 17, no. 2 (1998): 117–27; M.J. Rouault et al., "Mapping the Agulhas Current from Space: An Assessment of ASAR Surface Current Velocities," *Journal of Geophysical Research* 115 (2010): C10026; M.J. Rouault and P. Penven, "New Perspectives on Natal Pulses from Satellite Observations," *Journal of Geophysical Research* 116 (2011): C07013; B.C. Backeberg et al., "The Greater Agulhas Current System: An Integrated Study of Its Mesoscale Variability," *Journal of Operational Oceanography* 1, no. 1 (2008): 29–44; Morris et al., "The Importance of Monitoring the Greater Agulhas Current"; Michael A. Barnes and Christo Rautenbach, "Toward Operational Wave-Current Interactions over the Agulhas Current System," *Journal of Geophysical Research: Oceans* 125, no. 7 (2020): e2020JC016321. *Agulhas* is pronounced *ag-gull-lass* by English speakers, *ag-goo-yas* by Portuguese.

17. On the leakage of warm, salty Agulhas waters into the South Atlantic by rings, eddies, and filaments, see Olaf Boebel et al., "Path and Variability of the Agulhas Return Current," *Deep Sea Research Part II: Topical Studies in Oceanography* 50, no. 1 (2003): 35–56; Philip L. Richardson, "Agulhas Leakage into the Atlantic Estimated with Subsurface Floats and Surface Drifters," *Deep Sea Research Part I: Oceanographic Research Papers* 54, no. 8 (2007): 1361–89. The current is thought to influence the Atlantic Meridional Overturning Circulation on timescales from decades to millennia. Lisa M. Beal et al., "On the Role of the Agulhas System in Ocean Circulation and Climate," *Nature* 472, no. 7344 (2011): 429–36.

18. Both are western boundary currents, and hence swifter by definition than their eastern counterparts. The sources of speed differ. The Gulf Stream's speed is largely driven by wind stress and ocean depth, whereas the Agulhas's swiftness stems from the flow of the equatorial currents. Carl Wunsch, "What Is the Thermohaline Circulation?," *Science* 298, no. 5596 (2002): 1179–81; Lynne D. Talley et al., *Descriptive Physical Oceanography*, 6th ed. (Cambridge: Academic Press, 2011), chap. 9, sec. 3.2.

19. Wind speeds "exceeding 2 m/s 4.47 Mph (3.88 knots) are concentrated in the northern Agulhas Current with very few such speeds found either up- or downstream. The northern Agulhas Current has an exceptionally high mean speed and energy, but little variability, including seasonal variability. . . . Mean speeds lie between 1.4 m/s (3.13 Mph) 2.7 knots and 1.6 m/s (3.58 Mph) 3.11 knots whereas the peak speed from individual months is about 2.6 m/s (5.8 Mph) or 5 knots." Johann R.E. Lutjeharms, *The Agulhas Current* (Berlin: Springer, 2006): 93; Johann R.E. Lutjeharms et al., "The Hydrography and Water Masses of the Natal Bight, South Africa," *Continental Shelf Research* 20 (2000): 1907–39. The Gulf Stream's top speed is only slightly slower, at 5.6 miles per hour (nine kilometers per hour, or 4.86 knots).

20. The Agulhas Current "runs south along the east coast of southern Africa from ~25°S [Maputo] to 40°S. At an average rate of 0.5 to 1.4 miles (0.8 to 2.2 km) per hour reaching an estimated top speed of 5.8 miles (9.3 km) per hour off the southeast coast of South Africa. It is approximately 100 km wide and deep (2 km) meanders 10–15 km sideways on either side." Robert Todd et al., "Global Perspectives on Observing Ocean Boundary Current Systems," *Frontiers in Marine Science* 6 (2019): 423ff.; Lutjeharms, *The Agulhas Current;* Lisa M. Beal et al., "Capturing the Transport Variability of a Western Boundary Jet: Results from the Agulhas Current Time-Series Experiment (ACT)," *Journal of Physical Oceanography* 45 (2015): 1302–24; Lisa M. Beal and Harry L. Bryden, "The Velocity and Vorticity Structure of the Agulhas Current at 32°S," *Journal of Geophysical Research: Oceans* 104, no. C3 (1999): 5151–76.

The current continues both eastward and westward on either side of the African continent. On the return current to the Indian Ocean, see W.P.M. de Ruijter et al., "Indian-Atlantic Interocean Exchange: Dynamics, Estimation and Impact," *Journal of Geophysical Research: Oceans* 104, no. C9 (1999): 20885–910. On the leak into the South Atlantic, see Marten L. Gründlingh, "On the Course of the Agulhas Current," *South African Geographical Journal* 65, no. 1 (1983): 49–57; Marten L. Gründlingh and Alan F. Pearce, "Large Vortices in the Northern Agulhas Current," *Deep Sea Research Part A: Oceanographic Research Papers* 31, no. 9 (1984): 1149–56. One Sverdrup (Sv) equals one million cubic meters per second. The Agulhas is unusually deep, at places extending from the surface to the bottom of the ocean floor. This depth may be one of the reasons for its power.

21. "As correntes, a que nós havíamos grande medo [de] não nos deixar haver aquilo que desejávamos. E, daquele dia em diante, quis Deus que nós fôssemos avante e não a ré: praza a Ele que assim seja sempre." Velho, 30.

22. For some of the difficulties in identifying the Agulhas retroflection and leakage, see Y. Cheng et al., "Quantifying Agulhas Leakage in a High-Resolution Climate Model," *Journal of Climate* 29, no. 19 (2016): 6881–92; G.D. Quartly and M.A. Srokosz, "Seasonal Variations in the Region of the Agulhas Retroflection: Studies with Geosat and FRAM," *Journal of Physical Oceanography* 23 (1993): 2107–24.

23. Seventy years later, Manuel Perestrelo placed the start of the Natal coast at thirty-two degrees south. Manuel de Mesquita Perestrelo, *Carte de la côte d'Afrique* (s.n., 1915). The closest river is the Nenga.

24. Perestrelo's map places Pescaria at −29.3 degrees south, making Durban at −29.883 degrees the most likely location. See Perestrelo, "Demonstração da Costa de Boa Esperança até ao dos Correntes," 1576. Although the area northeast of Tugela Bluff has been proposed, it lacks the upwelling necessary for fishing. *Africa Pilot*, vol. 3, chap. 2, sec. 26.

25. The Haversine formula gives 198.18 nautical miles from Durban to Kosi Lake (−27.000325, 32.867813), the landmark nearest the start of the Agulhas Current at twenty-seven degrees south. A.L. Gordon, "Indian-Atlantic Transfer of Thermocline Water at the Agulhas Retroflection," *Science* 227, no. 4690 (1985): 1030–33.

26. The International Hydrographic Organization (IHO) defines the Mozambique Channel as the sea between the coasts of Mozambique and Madagascar,

extending the island's entire length from Cap d'Ambre to Cap Sainte-Marie. Cap Sainte-Marie lies twenty-six nautical miles north of the island's southern tip. "The E coast of Madagascar is, with good reason, noted for its inhospitable shore. The SE trade wind causes increasing strong winds and a heavy swell breaking on the coast from April to October. In summer, rain squalls obscure the land, making it hard to identify the landmarks; moreover, this is the season for cyclones, to which this coast is exposed. There are few good harbors to be found on this coast." *Africa Pilot*, vol. 3, chap. 10, sec. 1.

27. "The flow is dominated by a train of significant (diameters over 300 km) anti-cyclonic eddies that reach to the channel bottom. . . . [S]outhward propagation speed of the 25 eddies under examination was about 6 km/day between 12 S and 27 S in the Mozambique Channel, except for the region between 18 S and 21 S; where on average the southward propagation was only 3–4 km/day." M. W. Schouten et al., "An Oceanic Teleconnection Between the Equatorial and Southern Indian Ocean," *Geophysical Research Letters* 29, no. 16 (2002): 59-1–59-4. The authors add: "Most ocean models simulated the Mozambique Current as a weak, along-shelf, current with little variability. By contrast, modern satellite altimetric and other observations and eddy-permitting models indicate extraordinarily high variability along its purported trajectory. . . . The first research cruise designed specifically to investigate the . . . Mozambique Current . . . has revealed that no intense, coherent, western boundary layer-type current exists."

28. The greatest ship losses during the sixteenth century occurred either in the Agulhas region or in the Mozambique Channel itself. Paulo Guinote, "India Route Project: Ascensão e Declínio da Carreira da Índia," 37–38, https://alvarovelho.net/images/patrono/ASCENSAO_E_DECLINIO_DA_CARREIRADAINDIA.pdf. Guinote does not distinguish between the channel itself and the Agulhas Current regions.

29. "[A]ndámos tanto pelo mar, sem tomarmos porto, que não tínhamos já água que bebêssemos nem fazíamos já de comer senão com água salgada. E, para nosso beber, não nos davam senão um quartilho, de maneira que nos era necessário de tomarmos porto." Velho, 16–17.

30. Da Gama selected a second individual to accompany Afonso, but the text does not name him.

31. Velho, 15–16.

32. Velho, 16–17.

33. Velho, 16–17.

34. Their limited trade occurred with people from the interior, who acquired poor-quality cotton. The area lacked valuable trade goods, and in 1498 the people were unfamiliar with large, seagoing vessels. For the coastal mangrove distribution, see Aurelie Shapiro, *Mozambique Mangrove Extent 1995-Present* (Gland, Switzerland: 2018).

35. "E esta gente folgava muito; e nos traziam aos navios disso, que tinham em almadias, que eles têm. E nós isso mesmo íamos à sua aldeia a tomar água." Velho, 18. Chuabo (Echuwabo speakers) belong to the Macau group of Southern Bantu speakers.

36. Note that the fleet had already passed Sofala without any seagoing vessel in that harbor or at sea. By 1498, most of those engaged in the Arabian Sea trade resided farther north.

37. "E aqui nos adoeceram muitos homens, que lhes inchavam os pés e as mãos, e lhes cresciam as gengivas tanto sobre os dentes que os homens não podiam comer." Velho, 17–18.

38. "Esta terra é muito baixa e alagadiça e é de grandes arvoredos, os quais dão muitas frutas." Velho, 17–18. Marula trees grow up to sixty feet (eighteen meters) tall. Kei apple trees reach twenty to thirty feet (eight to nine meters) in moist open woodland. Natal plums grow on shrubs. The National Academy of Sciences reports: "All African fruits are useful sources of nutrients, particularly vitamin C." National Research Council, *Lost Crops of Africa* (Washington, DC: National Academies Press, 2008), vol. 3, 8.

39. Velho, 21.

40. "[N]os parecia que nos íamos chegando para onde desejávamos." Velho, 18.

41. Velho, 23–24.

42. Velho, 23–24. Albuquerque agrees that tides were responsible for Coelho's temporary stranding. See Luís de Albuquerque, *A Viagem de Vasco da Gama entre Moçambique and Melinde, segundo Os Luíadas e segundo as Crónicas* (Lisbon: Junta de Investigações do Ultramar, 1972), 33.

43. Velho, 20.

44. Velho, 21.

45. "Keel stitched with coconut fiber matting sail made of plaid strips of the leaflets of the mkoma palm." James Hornell, "The Sea-Going Mtepe and Dáu of the Lamu Archipelago," *Mariners Mirror* 27, no. 1 (1941): 55–58.

46. Velho, 23.

47. Velho, 22.

48. Caliph Abd al-Malik ibn Marwan issued the gold dinar in AH 77 (696–97 CE). The weight of the dinar is one mithqal (4.25 grams). Velho, 26.

49. Velho, 27.

50. "Neste mesmo dia [abril 14] ao sol-posto lançámos âncora em frente dum lugar que se chama Melinde, o qual está de Mombaça 30 léguas. E, de Mombaça a esta vila há estes lugares: Benapa, e Toça e Nuguo-Quioniete." Velho, 26.

51. R. Raoul McLaughlin, *The Roman Empire and the Indian Ocean: The Ancient World Economy and the Kingdoms of Africa, Arabia and India*, ed. P. De Souza et al. (Barnsley, UK: Pen and Sword Books, 2018); *The Sea in History—The Ancient World*, vol. 1, ed. Philip de Souza et al. (UK: Boydell & Brewer, 2017); Eivind Seland, "The Persian Gulf or the Red Sea? Two Axes in Ancient Indian Ocean Trade, Where to Go and Why," *World Archaeology* 43, no. 3 (2011): 398–409.

52. Luís Filipe Thomaz, "Les langues et la communication dans le Mare Luso-Indicum," *Nuevo Mundo Mundos Nuevos*, 2013, http://journals.openedition.org/nuevomundo/65059.

53. Velho, 65–71, 64, 82–83, 85.

13. A DREADFUL MISTAKE

1. Kevin E. Trenberth et al., "The Global Monsoon as Seen Through the Divergent Atmospheric Circulation," *Journal of Climate* 13, no. 22 (2000): 3969–93; Edward Pollard and Okeny Charles Kinyera, "The Swahili Coast and the Indian Ocean Trade Patterns in the 7th–10th Centuries CE," *Journal of Southern African Studies* 43, no. 5 (2017): 927–47; Blanche Rocha D'Souza, *Harnessing the Trade Winds: The Story of the Centuries-Old Indian Trade with East Africa, Using the Monsoon Winds* (Nairobi: Zand Graphics, 2008).

2. Velho, 87, 94.

3. "Indo nós obra de 200 léguas, em pego, donde partíramos, disse o mouro que tomáramos que já lhe parecia tempo para não encobrir nada, que era verdade que, estando ele em casa do seu senhor, lhe vieram dizer como nós andávamos perdidos ao longo da costa; que nós não sabíamos tornar para [a] nossa terra." Velho, 76.

4. His Venetian and Genoese counterparts dispatched their goods on ships owned by Arab traders. Within weeks, their merchandise reached the safety of the Gulfs of Oman and Aden and the Red Sea's entrance. They were transported across land from the northern Red Sea to the Mediterranean.

5. Velho, 76. But when da Gama was offered a significant sum for the remaining Arabian hulk, he turned it down, saying that since it was an enemy ship, he would burn it instead.

6. Velho, 67, 71.

7. From December to March (the northeast monsoon), the Somali Current flows southward to two to three degrees south, which would have been excellent for da Gama. Where the East African Coastal Current meets the Somali Current, the currents turn eastward into the Indian Ocean, forming the South Equatorial Countercurrent. Walter Düing and Friedrich Schott, "Measurements in the Source Region of the Somali Current during the Monsoon Reversal," *Journal of Physical Oceanography* 8, no. 2 (1978): 278–89. In Kenya and Somalia, the dominant wind between June and October is mainly from the southwest.

8. Groundwater could be found in such places as Salah, Sawqirah, and Sharbithat Ash Shuwaymiyyah in Oman and Qishn in Yemen. For the groundwater springs along the coast of Oman, see Rachel McDonnell, "Groundwater Use and Policies in Oman," *International Water Management Group Project Report* 14 (2016); Düing and Schott, 1978; Fanack Water, 2019, "Water Resources in Yemen," https://water.fanack.com/yemen/water-resources-yemen/.

9. Using a great circle route and the shortest possible distance yields 1,984.88 nautical miles (Haversine calculation).

10. The "Great Whirl," a historically significant northern gyre in the "two gyre" system, often occurs between five and nine degrees north. Melzer and colleagues found that the Great Whirl lasted an average of 198 days between 1995 and 2015 and traveled back and forth for 4,100 kilometers. The average peak surface radius was 309 kilometers. B. A. Melzer et al., "Evolution of the Great Whirl Using an Altimetry-Based Eddy Tracking Algorithm," *Geophysical Research Letters* 46, no. 8 (2019): 4378–85. The historical significance is further explored in the following works: He Wang et al., "Seasonal Cycle and Annual Reversal of the Somali Current in an Eddy-Resolving Global Ocean

Model," *Journal of Geophysical Research: Oceans* 23, no. 9 (2018): 102374; Stuart C. Painter, "The Biogeochemistry and Oceanography of the East African Coastal Current," *Progress in Oceanography* 186 (2020): 102374; Trenberth et al., "The Global Monsoon as Seen Through the Divergent Atmospheric Circulation"; Friedrich A. Schott and Jüirgen Fisch, "Winter Monsoon Circulation of the Northern Arabian Sea and Somali Current," *Journal of Geophysical Research: Oceans* 105, no. C3 (2000): 6359–637; Friedrich Schott et al., "The Somali Current at the Equator: Annual Cycle of Currents and Transports in the Upper 1000 m and Connection to Neighbouring Latitudes," *Deep Sea Research A: Oceanographic Research Papers* 37, no. 12 (1990): 1825–48; Robert L. Molinari et al., "Surface Current Distributions in the Tropical Indian Ocean Derived from Compilations of Surface Buoy Trajectories," *Journal of Geophysical Research* 9 (1990): 7217–38.

11. The summer Somali Current, a key component of my research, to the north of two to three degrees south, gradually breaks up during the fall transition period and is replaced by the southwestward winter Somali Current with the onset of the northeast monsoon in December. These currents, particularly the Somali Current, play a significant role in shaping sailing along the western edge of the Arabian sea. Pollard and Kinyera, "The Swahili Coast and the Indian Ocean Trade," 927–47.

12. The text does not give an exact date. It only states that it was three days short of three months. Thirty days in November plus thirty-one in October and December equals ninety-two days minus three, which is eighty-nine days. If the starting date is included, then ninety days would have passed.

13. James Lind innovated the controlled clinical trial for scurvy remedies, but da Gama knew of the appropriate treatment 260 years earlier. James Lind, *A Treatise on the Scurvy in Three Parts*, 2nd edition (London: A. Millar, 1757). Johann Friedrich Bachstrom and Coenraad Wishoff, *Observationes circa scorbutum, ejusque indolem, causas, signa, et curam, institutae: eorum praeprimis in usum, qui groenlandiam & indiam orientis petunt* (Leiden, The Netherlands: Conradum Wishof, 1734); Thomas Jukes, "Historical Perspectives: The Prevention and Conquest of Scurvy, Beri-Beri, and Pellagra, *Preventative Medicine* 18 (1989), 877. A clinical description of severe scurvy appears in Velho's account: "No's adoeceu toda a gente das gengivas, que lhes cresciam sobre os dentes em tal maneira que não podiam comer; e isso mesmo lhes inchavam as pernas, e [com] grandes outros incha ços pelo corpo, de guisa que lavravam [por] um homem tanto que morria sem ter outra nenhuma doença." Velho, 17–18.

14. Fontoura da Costa inserts an apostrophe in *Mtang'ata* (the regional Swahili dialect rather than the name of a place). Anonymous, "Regimentos do Cruzeiro do Sul e de diversas estrelas," in A. Fontoura da [Abel Fontoura] Costa, *A marinharia dos descobrimentos*, 3rd edition, 118–47 (Lisbon: Agência Geral do Ultramar, 1960). Axelson identifies the place, without the apostrophe, as Mtangata Reef. Eric Axelson, *Portuguese in South-East Africa 1488–1600* (Cape Town: Struik, 1973), 63 n54. Da Gama named the reefs Baixos de São Jorge. Little information is provided about the remaining losses. One African deserted to the Tsonga realm, and an unknown number of convicts were deliberately left behind in areas where the Portuguese planned to establish

commercial relationships. The remainder were fatally injured on board or perished from other illnesses.

15. Velho, January 13, 1499, n256. The actual location appears to be the "Baixos de São Rafael," identified on the Cantino map of 1502 and possibly the reefs off Karange Island. Sir William Snow Harris published the first systematic study of lightning strikes on wooden ships, using data from 235 strikes on British Navy vessels from 1793 to 1839. W.S. Harris and R.B. Forbes, *Protection of Ships from Lightning, According to Principles Established by Sir W.S. Harris, F.R.S., and Approved After Eighteen Years' Experience in the British Navy. . . .* (s.n., 1848).

16. This obscure location hid the damaged vessels from potential attackers because it did did not form part of the chain of Arabian Sea trading posts and hence held only local fishing boats. Da Gama named the reef the São Rafael (Baixos de São Rafael) for the ship they had burned on the reefs.

17. São Jorge or Goa Island is a small coral island at the entrance to Mozambique harbor. Boats enter the harbor on either side of this island. Da Gama arrived on February 1, 1499, on his return journey from India.

18. "Bem mortos do frio das grandes brisas." Velho, 82.

19. Castanheda reports that only fifty-five of the initial 145 crew members returned to Lisbon—in other words, 63 percent of the crew were lost. However, only a third of the losses could be attributed to scurvy, the majority to other causes. There was a single desertion and five were lost to accidents. Only 20 percent of the crew were lost to scurvy. Fernão Lopes de Castanheda, *Primeiro livro da historia do descobrimento e conquista da india* (Lisbon: J. de Barreyra e J. Alvarez, 1552).

20. Eight weeks and five days.

14. THE SALTY TEARS OF THE ATLANTIC

1. This chapter's epigraph, the beginning of Fernando Pessoa's poem "Mar Português," ends with "of Portugal," but the stanza stands perfectly well without this phrase. "Ó mar salgado, quanto do teu sal / São lágrimas de Portugal." Translation by the author.

2. These routes stand out in any software that tracks maritime travel in real time. At the time of this writing, even the accessible version of the Marine Traffic app illustrates these patterns strikingly.

3. Magellan would have needed to understand the Coriolis effect to transition from the South to the North Pacific, where he died. On the absence of pilots' names, see Francisco Contente Domingues, *Os navios do mar oceano: teoria e empiria na arquitetura naval portuguesa dos séculos XVI e XVII* (Lisbon: Centro de História da Universidade de Lisboa, 2004).

4. "Habet item rex in curia sua multos filios dominorum ex Aethiopia, quos informat ritus et legem mostram." Frederick Kunstmann, *Hieronymus Münzer's Bericht über die Entdeckung der Guinea* (Berlin: Verlag der K. Akadamie, 1854), 357.

5. "Et mittit eis continuo munera, quibus placat eos ita quod sui securi jam plures terras in continenti Aethiopiae peragrare possunt et singula inquirere,

nam olim solum circa maritimam Aethiopiam negotiatus fuit, nec sui terram praesumebant ad multa milliaria ingredi. Habet item . . . de vario idiomate, qui linguam portugalensem sciunt, quia várias linguas habent, et bis interpretibus usus quasi totam Aethiopiam superambulat." Hieronymus Münzer, "Do descobrimento da África marítima e ocidental, isto é, da Guiné pelo infante D Henrique de Portugal in Jerónimo Munzer," in *As Viagens dos descobrimentos*, ed. José Manuel Garcia, 55–71 (Lisbon: Editorial Presença, 1983 [1495]), 357.

6. "Habet item . . . de vario idiomate, qui linguam portugalensem sciunt, quia várias linguas habent, et bis interpretibus usus quasi totam Aethiopiam superambulat." Münzer, 357.

7. "[M]ajores reges continuo muneribus placat, non enim possibile est, ipsuni eis dominari. Etiam si dominium haberet, parum commodi sibi inde esset." Münzer, 357.

Bibliography

HISTORICAL PRIMARY SOURCES

Albuquerque, Luís de, and André Pires. *O Livro de marinharia de André Pires.* 2nd ed. Lisbon: Vega, 1989.

Albuquerque, Luís de, Maria Emília Madeira Santos, and Maria Luísa Oliveira Esteves. *Portugaliae monumenta Africana.* Lisbon: Instituto de Investigação Científica Tropical, Esteves, Centro de Estudos de História e Cartografia Antiga, Comissão Nacional para as Comemorações dos Decobrimentos Portugueses, 1993.

Alfonso IX and Gregorio López. *Las siete partidas del sabio rey don Alonso el Nono.* Madrid: Boletín Oficial del Estado, 1985.

Anonymous. Cantino chart. Modena, Italy: Biblioteca Estense, 1502.

Anonymous. Map of Africa. Munich: Bayerische Staatsbibliothek, 1504.

Anonymous. "Regimentos do Cruzeiro do Sul e de diversas estrelas." In A. Fontoura da [Abel Fontoura] Costa, *A marinharia dos descobrimentos,* 118–47. 3rd ed. Lisbon: Agência Geral do Ultramar, 1960.

Aristotle. *Meteorologica.* Translated by Henry Desmond Pritchard Lee. Cambridge, MA: Harvard University Press, 1952.

Barros, João de, and Diogo de Couto. *Décadas da Ásia, vol. 1.* Lisbon: Na Regia officina typografica, 1777.

Behaim, Martin. "Erdapfel 1492." Globe. Nuremberg, 1490–92.

Bosman, Willem, and G. Brown Goode. *A New and Accurate Description of the Coast of Guinea, Divided into the Gold, the Slave, and the Ivory Coasts.* London: James Knapton, 1705.

Bosman, Willem, Anthony Schouten, Jan Snoek, and David van Nyendael. *Nauwkeurige Beschryving Van De Guinese Goud- Tand- En Slave-kust.* Utrecht: Anthony Schouten, 1704.

Brásio, António. *Monumenta missionaria Africana. Africa ocidental, 1685–1699 e suplemento aos séculos XV, XVI, XVII.* Lisbon: Agência Geral do Ultramar, Divisão de Publicações e Biblioteca, 1958.

Cadamosto, Luís de, and Pedro de Sintra. *Viagens de Luís de Cadamosto e de Pedro de Sintra.* Lisbon: Academia Portuguesa de História, 1948.

"Capitulações dos reis de Castela para o contrato de casamento de D. Afonso V (1470–1472)." Translated by Diana Martins. *Fragmenta Histórica—História, Paleografia e Diplomática* 6 (2018): 153–55.

Castanheda, Fernão Lopes de. *Primeiro livro da historia do descobrimento e conquista da india.* Lisbon: J. de Barreyra e J. Alvarez, 1552.

Castilho, Alexandre Magno de. *Descripção e roteiro da costa occidental de Africa desde o cabo de Espartel até o das Agulhas.* Lisbon: Imprensa Nacional, 1866.

Cervantes Saavedra, Miguel de. *Los Trabajos de Persiles y Sigismunda.* Edited by Laura Fernández, Ignacio García Aguilar, Carlos Romero Muñoz, and Isabel Lozano-Renieblas. Madrid: Real Academia Española, 2017.

Cervantes Saavedra, Miguel de. *Trato de Argel.* Barcelona, Spain: Linkgua Ediciones, 2011.

Cervantes Saavedra, Miguel de, and Juan Bautista Avalle-Arce. *La Galatea.* Madrid: Espasa-Calpe, 1961.

Coelho, José Ramo, ed. *Alguns documentos do Archivo Nacional da Torre do Tombo ácerca das navegações e conquistas portuguezas publicados por ordem do governo de sua majestade fidelissima ao celebrar-se a commemoração quadricentenaria do descobrimento da America.* Lisbon: Imprensa Nacional, 1892.

Coleridge, Samuel Taylor. *The Rime of the Ancient Mariner.* Boston: Estes and Lauriat, 1884.

Columbus, Christopher. *Mappa Mundi*, no date, Bibliothèque nationale de France, Paris.

Cunliffe, Barry W. *The Extraordinary Voyage of Pytheas the Greek.* London: Allen Lane, 2001.

d'Almada, André Alvares [active 1594], and António Brásio. *Tratado Breve Dos Rios de Guiné Do Cabo Verde Dês Do Rio Sanagá Até Os Baixos de Santa Ana de Todas As Naçoes de Negros Que Há Na Dita Costa e de Seus Costumes, Armas, Trajos, Juramentos, Guerras. Ano 1594.* Lisbon: Editorial L.I.A.M., 1654.

Dapper, Olfert. *Naukeurige beschrijvinge der Afrikaensche gewesten van Egypt, Barbaryen, Libyen, Biledulgerid, Negroslant, Guinea, Ethiopiën, Abyssinie . . . getrokken uit verscheyde hedendaegse lantbeschrijvers en geschriften van bereisde ondersoekers dier landen.* Amsterdam: Jacob van Meurs, 1668.

Donelha, André. *Descrição da Serra Leoa e dos rios de Guiné do Cabo Verde (1625).* Lisbon: Junta de Investigações Científicas do Ultramar, 1977.

Egerton Collection. British Library, London.

Eratosthenes. *Eratosthenes' Geography.* Edited and translated by Duane W. Roller. Princeton, NJ: Princeton University Press, 2010.

Fernandes, Valentim, José Pereira da Costa, and V. Fernandes. *Códice Valentim Fernandes*, ed. Joaquim Bensaúde; leitura paleográfica, notas e índice pelo académico de número, José Pereira da Costa. Lisbon: Academia Portuguesa da História, 1997.

Fernández de Palencia, Alfonso, and Antonio Paz y Meliá. *Crónica de Enrique IV*. Madrid: Ediciones Atlas, 1973.

Findlay, Alexander George. *A Directory for the North Atlantic Ocean: Comprising Instructions General and Particular for Its Navigation*. Edited by W.R. Kettle. 16th ed. London: Imray, Laurie, Norie & Wilson, 1918.

Godinho, Vitorino Magalhães, ed. *Documentos sôbre a expansão portuguesa*. 3 vols. Lisbon: Editorial "Gleba," 1943.

Góis, Damião de. *Chronica do serenissimo senhor rei D. Emanuel*. Coimbra, Portugal: Na real officina da Universidade, 1790.

Goldsmiths'-Kress Library of Economic Literature, London.

Harris, W. S., and R. B. Forbes. *Protection of Ships from Lightning, According to Principles Established by Sir W. S. Harris, F.R.S., and Approved After Eighteen Years' Experience in the British Navy*. . . . S.n., 1848.

Herodotus. *Herodotus*. Translated by William Beloe. 4th ed. 4 vols. London: F.C. and J. Rivington, 1821.

Homer. *The Odyssey*. Translated by Emily Wilson. New York: W.W. Norton.

Hondius, Jodocus. "Anzicana . . . populi antropophagi." Map. In *Nova Africae tabula*. Amsterdam: Hondius, 1606.

Hopkins, J.F.P., and Nehemia Levtzion, eds. and trans. *Corpus of Early Arabic Sources for West African History*. Princeton, NJ: Markus Wiener.

Idrīsī, Muḥammad. *Geographie, d'Edrisi*. Translated by Amedée Jaubert. 2 vols. Paris: Imprimerie Royale, 1840.

Jernside, Bjørn. *The Saga of the Volsungs, with the Saga of Ragnar Lothbrok*. Translated and with an introduction by Jackson Crawford. Indianapolis, IN: Hackett, 2017.

La Fosse, Eustache de. *Voyage d'Eustache Delafosse sur la côte de Guinée, au Portugal et en Espagne (1479–1481)*. Edited by Theodore Monod and Denis Escudier. Paris: Editions Chandeigne, 1992.

Lavanha, João Baptista. *Regimento Nautico*. Lisbon: Em casa de Simão Lopez, 1595.

Li, Andrés de, and Valentim Fernandes. *Reportório dos tempos: tresladado de Castelhano em Portuguez per Valentim Fernandez. Com o regimento da declinação do sol . . . tirada pontualmente do Zacuto pelo honrado Gaspar Nicolas*. . . . Reproduction facsimilé de l'exemplaire appartenant à la Bibliotheque nationale de Lisbonne, édition 1563.

Livro do Armeiro-Mor. Lisbon, n.p., 1509.

Lopes da Costa Almeida, Antonio. *Roteiro geral dos mares, costas, ilhas, e baixos reconhecidos no globo*. 14 vols. Lisbon: Tipografia de Academia, 1835–49.

Lucena, Vasco Fernandes de. *The Obedience of a King of Portugal*. Translated and with commentary by Francis M. Rogers. Minneapolis: University of Minnesota Press, 1958.

Manitius, Carl 1894. *Hipparchi in Arati et Eudoxi Phaenomena commentariorum libri tres*. Leipzig, Germany: B.G. Teubner, 1894.

Marees, Pieter de, and Samuel Pierre L'Honoré Naber. *Beschryvinghe ende historische verhael van het Gout koninekrijek van Gunea anders de Gout Custe de Mina genaemt liggende in het deel van Africa*. The Hague: M. Nijhoff, 1912 [1602].

Mas'udi, Muruj al-. *L'Abrégé des Merveilles Traduit de l'arabe d'après les manuscrits de la Bibliothèque nationale de Paris par le Bernard Carra de Vaux*. Paris: n.p., 1898.

Maury, Matthew Fontaine. "Maury's Wind & Current Chart, [North Atlantic]: No. 2, Series A." Washington, DC: U.S. Hydrographic Office, 1851.

Mueller, Wilhelm Johan. *Die Africanische auff der Guineischen Gold-Cust gelegene Landschafft Fetu*. Hamburg, 1673.

Münzer, Hieronymus. "Do descobrimento da África marítima e ocidental, isto é, da Guiné pelo infante D Henrique de Portugal in Jerónimo Munzer." In *As Viagens dos descobrimentos*, edited by José Manuel Garcia, 55–71. Lisbon: Editorial Presença, 1983 [1495].

Nebrija, Antonio de. *Gramática Castellana*. Halle, Belgium: Max Niemeyer, 1909.

Nicholson, Sharon E. *Dryland Climatology*. Cambridge: Cambridge University Press, 2011.

Pacheco Pereira, Duarte, and Joaquim Barradas de Carvalho. *Esmeraldo de situ orbis de Duarte Pacheco Pereira / édition critique et commentée [par] Joaquim Barradas de Carvalho*. Lisbon: Fundação Calouste Gulbenkian, Serviço de Educação, 1992 [1892].

Palencia, Alfonso Fernández de [1423–92]. *Crónica de Enrique IV*. 5 vols. Madrid: [Tip. de la Revista de archivos], 1904.

Pessoa, Fernando. *Mensagem*. Lisbon: Babel, 2010.

Pina, Rui de. *Chronica Deel-rei D. Affonso V*. 3 vols. Lisbon: Escriptorio, 1901–2.

Pina, Rui de. *Chronica d'El-Rei D. Duarte*. Edited by Gabriel Pereira. Lisbon: Escriptorio, 1901.

Pulgar, Fernando del. *Crónica de los Reyes Católicos* [1481–1490]. Edited by Juan de Mata Carriazo. Madrid: Espasa-Calpe, 1943.

Qalqashandī, Aḥmad ibn 'Alī, Ferdinand Wüstenfeld, Bernard Michel, et al. *Two Studies on the Geography and Administration of Egypt According to al-Qalqašandī (d. 1418)*, edited by Fuat Sezgin Mazen Amawi, Carl Ehrig-Eggert, and Eckhard Neubauer. Frankfurt: Institute for the History of Arabic-Islamic Science, Johann Wolfgang Goethe University, 1992.

Radulet, Carmen M. *O cronista Rui de Pina e a "Relação do Reino do Congo": manuscrito inédito do "Códice Riccardiano 1910."* Lisbon: Comissão Nacional para as Comemorações dos Descobrimentos Portugueses, 1992.

Ramin, Jacques. *Le Périple d'Hannon / The Periplus of Hanno*. Oxford: British Archaeological Reports, 1976.

Resende, Garcia de, and Gabriel Pereira. *Chronica De El-rei D. João II*. Lisbon: Escriptorio, 1902.

Robertson, John, Philip Carteret Webb, Swithin Adee, et al. "An Extraordinary and Surprising Agitation of the Waters, Though Without Any Perceptible Motion of the Earth." *Philosophical Transactions of the Royal Society of London* 49 (1755): 351–98.

Sailing Directions for the East-India or Oriental Pilot, for the Navigation Between England and the Cape of Good Hope. London: R. Sayer and J. Bennett, 1781.

Scott, Lionel. *Pytheas of Massalia Texts, Translation, and Commentary*. New York: Routledge, 2022.

Seneca, Lucius Annaeus. *The Suasoriae of Seneca the Elder; Introductory Essay, Text, Translation and Explanatory Notes; Being the "Liber Suasoriarum" of the Work Entitled L. Annaei Senecae Oratorum et Rhetorum, Sententiae, Divisiones, Colores*. Edited and translated by William A. Edward. London: University Press, 1928.

Service hydrographique de la marine. Instructions nautiques: Océan Atlantique est, Afrique, côte ouest, vol. 2. Du Cap des Palmes au Cap des Aiguilles. Paris: Imprimerie nationale, 1900.

Sousa, João de. *Documentos Arábicos para a história portuguesa*. Lisbon: Academia Real das Ciências, 1790.

Thucydides. *History of the Peloponnesian War, Books I–VIII*. Translated by Charles Forster Smith. 4 vols. Cambridge, MA: Harvard University, 1980.

UK Hydrographic Office. *The African Pilot or Sailing Directions for the Western Coast of Africa, vol. 1*. London: UK Hydrographic Office, 1856.

UK Hydrographic Office. *Africa Pilot, vol. 2: Comprising the West Coast of Africa from Bakasi Peninsula to Cape Agulhas, Islands in the Bight of Biafra, Ascension Island, Saint Helena Island, Tristan da Cunha Group, and Gough Island*. Taunton: UK Hydrographic Office, 2022.

UK Hydrographic Office. *Africa Pilot, vol. 3: South and East Coasts of Africa from Cape Agulhas to Ras Binnah, Including the Islands of Zanzibar and Pemba*. Taunton: UK Hydrographic Office, 2022.

U.S. Hydrographic Office. *Atlas of Pilot Charts North Atlantic Ocean (Including the Gulf of Mexico)*. Publication 106. Washington, DC: U.S. Hydrographic Office, 2002.

U.S. Hydrographic Office. *Atlas of Pilot Charts South Atlantic Ocean*. Publication 105. Washington, DC: U.S. Hydrographic Office, 2009.

U.S. Hydrographic Office. *Enroute (SW Coast of Africa)*. Washington, DC: U.S. Hydrographic Office, 2021.

U.S. Hydrographic Office. *Sailing Directions*. Washington, DC: U.S. Hydrographic Office, 1916.

U.S. Hydrographic Office. *Sailing Directions Enroute*. Washington, DC: U.S. Hydrographic Office, 2019.

U.S. Hydrographic Office. *Sailing Directions (Enroute) West Coast of Europe and Northwest Africa*. Publication 143. Washington, DC: U.S. Hydrographic Office, 2025.

U.S. Hydrographic Office. *Sailing Directions for the Southeast Coast of Africa*. 3rd ed. Washington, DC: U.S. Hydrographic Office, 1936.

U.S. Hydrographic Office. *Sailing Directions for the West Coasts of Spain, Portugal, and Northwest*. Washington, DC: U.S. Hydrographic Office, 1908.

Usodimare, Antoniotto. "Carta de Antoniotto Usodimare." In *Viagens dos descobrimentos*, edited by José M. Garcia, 139–44. Lisbon: Editorial Presença, 1983.

Velho, Alvaro. *Roteiro da primeira viagem de Vasco Da Gama*. Edited by Neves Águas. Mem Martins, Portugal: Publicacões Europa-América, n.d.

Velho, Alvaro. *Roteiro da primeira viagem de Vasco de Gama à Índia, 1497–1499 = Journal of the First Voyage of Vasco da Gama to India, 1497–1499*.

Edited by Luís Fernando de Sá Fardilha and Maria de Lurdes Correia Fernandes. Porto, Portugal: U. Porto Edições, 2016.

Velho, Alvaro, Gago Coutinho, and Franz Hümmerich. *Diário da viagem de Vasco da Gama; / facsímile do códice original, transcrição e versão em grafia actualizada, com uma introd. por Damião Peres. Leitura paleográfica por António Baião e A. de Magalhães Basto. Texto actualizado por A. de Magalhães Basto.* Porto, Portugal: Livraria Civilização, 1945.

Velho, Alvaro, and Abel Fontoura da Costa. *Roteiro da primeira viagem de Vasco da Gama (1497–1499)*, with preface and notes by A. Fontoura da Costa. 3rd ed. Lisbon: Agência-Geral do Ultramar, 1969.

Virgil. *Georgics*. Translated by Peter Fallon. Introduction and notes by Elaine Fantham. Oxford: Oxford University Press, 2006.

Whalley, Joyce Irene. *Pliny the Elder, Historia naturalis*. London: Sidwick & Jackson, 1982.

Zacuto, Abraham ben Samuel. *O Regimento da declinação do sol . . . tirada . . . do Zacuto pelo honrado Gaspar Nicolas, mestre sufficiente nesta arte* [1563]. Geneva, Switzerland: [Société Sadag], 1917.

Zacuto, Abraham ben Samuel, and Joseph Vecinho. *Almanach Perpetuum Celestium Motuum (Radix 1473): Tabulae Astronomicae*. Leiria, Portugal: J.B. Obernetter, 1496.

Zurara, Gomes Eanes de, and Torquato de Sousa Soares. *Crónica dos feitos notáveis que se passaram na conquista da Guiné por mandado do Infante D. Henrique*. Lisbon: Academia Portuguesa da História, 1978.

OTHER SOURCES

Adiele, Pius Onyemechi. *The Popes, the Catholic Church and the Transatlantic Enslavement of Black Africans 1418–1839*. Hildesheim, Germany: Georg Olms, 2017.

Albuquerque, Luís de. *Ciência e Experiencia nos Descobrimlentos Portugueses*. Lisbon: Instituto de Cultura e Língua Portugues, 1983.

Albuquerque, Luís de. *Curso de História da Náutica*. Coimbra, Portugal: Livraria Almedina, 1972.

Allen, Martin. "Medieval Merchants and the English Mints and Exchanges, 973–1489." In *Medieval Merchants and Money: Essays in Honour of James L. Bolton*, edited by Martin Allen and Matthew Davies, 197–212. London: University of London Press, 2016.

Andersen, H., J. Cermak, I. Solodovnik, et al. "Spatiotemporal Dynamics of Fog and Low Clouds in the Namib Unveiled with Ground- and Space-Based Observations." *Atmospheric Chemistry and Physics* 19 (2019): 4383–92.

Appiah, Kwame Anthony, and Henry Louis Gates Jr. *Encyclopedia of Africa*. Oxford: Oxford University Press, 2010.

Arístegui, Javier, Xosé A. Álvarez-Salgado, Eric D. Barton, et al. "Oceanography and Fisheries of the Canary Current/Iberian Region of the Eastern North Atlantic." In *The Sea, vol. 14*, edited by Allan R. Robinson and Kenneth H. Brink. Cambridge, MA: Harvard University Press, 2004.

Ash, Eric H. *Power, Knowledge, and Expertise in Elizabethan England*. Baltimore: Johns Hopkins University Press, 2004.

Aubynn, Anthony. "Sustainable Solution or a Marriage of Inconvenience? The Coexistence of Large-Scale and Artisanal and Small-Scale Mining on the Abosso Goldfields Concession in Western Ghana." *Resources Policy* 34, no. 1 (2009): 64–70.

Axelson, Eric. *Portuguese in South-East Africa 1488–1600*. Cape Town: Struik 1973.

Bachstrom, Johann Friedrich, and Coenraad Wishoff. *Observationes circa scorbutum, ejusque indolem, causas, signa, et curam, institutae: eorum praeprimis in usum, qui groenlandiam & indiam orientis petunt*. Leiden, The Netherlands: Conradum Wishof, 1734.

Barfod, Dan N., and J. Godfrey Fitton. "Pleistocene Volcanism on São Tomé, Gulf of Guinea, West Africa." *Quaternary Geochronology* 21 (2014): 77–89.

Barnier, Bernard, Thierry Reynaud, Aike Beckmann, Claus Böning, Jean-Marc Molines, Sally Barnard, and Yanli Jia. "On the Seasonal Variability and Eddies in the North Brazil Current: Insights from Model Intercomparison Experiments." *Progress in Oceanography* 48, no. 2–3 (2001): 195–230.

Barros, Maria Filomena Lopes de. "A minoria muçulmana do reino português e os contactos diplomáticos com o dār al-islām." In *Comunicación y política en la baja Edad Media*, edited by Neéstor Vigil Montes. Évora, Portugal: Cidehus, 2019.

Barry, Roger Graham, and Richard J. Chorley. *Atmosphere, Weather, and Climate*. London: Routledge, 1992.

Basu, Niladri, Edith Clarke, Allyson Green, et al. "Integrated Assessment of Artisanal and Small-Scale Gold Mining in Ghana—Part 1: Human Health." *International Journal of Environmental Ressearch and Public Health* 12, no. 5 (2015): 5143–76.

BBC News. "Women Die in Ghana Mine Collapse." November 12, 2009.

Binet, D. "Climate and Pelagic Fisheries in the Canary and Guinea Currents 1964–1993: The Role of Trade Winds and the Southern Oscillation." *Oceanologica Acta* 20, no. 1 (1997): 177–90.

Binet, Denis, Bertrand Gobert, and Lucien Maloueki. "El Niño–like Warm Events in the Eastern Atlantic (6°N, 20°S) and Fish Availability from Congo to Angola (1964–1999)." *Aquatic Living Resources* 14 (2001): 99–113.

Blanco-Chao, Ramon, Manuela Costa-Casis, Daniel Cajade-Pascual, and Gonzalo Gómez-Rey. "Coastal Retreat and Sedimentation During the Last 3000 Years. Atlantic Coast of NW Spain." *Journal of Marine Science & Engineering* 7, no. 10 (2019): art. 331.

Boahn, A. A. "The States and Cultures of the Lower Guinean Coast." In *General History of Africa*, abridged ed., vol. 5: *Africa from the Sixteenth to the Eighteenth Century*, edited by Bethwell A. Ogot, 399–433. Paris: UNESCO, 1992.

Bochaca, Michel. *Voyager par mer au Moyen Âge: Navigations dans le Ponant au XVe Siècle*. Paris: Editions Gisserot, 2023.

Bosson, Kouacou, Angora Aman, Elisée Toualy, and Sabine Arnault. "The Surface Guinea Current Variability from Satellite Data." *Regional Studies in Marine Science* 64 (2023): 2352–4855.

Bouchon, Geneviève. *Vasco de Gama*. Translated by Ana Santos Silva, Maria da Graça Pinhão, Manuela Torres, and Joaquim A. Nogueira Gil. Lisbon: Terramar, 1998.

Brito, Renata. "9 Bodies Found Adrift in a Boat off Brazil Were Likely Migrants from Mauritania and Mali, Police Say." Associated Press, April 16, 2024. https://apnews.com/article/brazil-boat-african-migration-mauritania-mali-ea8303f76c3699bc6b98d9730bb5215b.

Brito, Renata, and Felipe Dana. "Adrift in the Atlantic, a Boat of Death and Lost Dreams." Associated Press, April 11, 2023.

Buckley, Martha W., and John Marshall. "Observations, Inferences, and Mechanisms of the Atlantic Meridional Overturning Circulation: A Review." *Reviews of Geophysics* 54, no. 1 (2016): 5–63.

Cairncross, Bruce. "The Where of Mineral Names: Namibite, Namib Desert, Copper Valley, Khorixas District and Region Namibia." *Rocks & Minerals* 93 (2018): 184–87.

Catewicz, Zygmunt, and Ryszard Siwecki. "On Currents in the Coastal Zone of African Shelf off Saint Louis." *Oceanologia* 21 (1985): 59–75.

Cereira, João Cordeiro. "Resgate do ouro na Costa da Mina nos reinados de D. João III e D. Sebastião." *Studia: revista semestral* 50 (1991): 5–48.

Chabás, José, and Bernard Cohen. "Abraham Zacut and the Transition from Manuscript to Print, Astronomy in the Iberian Peninsula." *Transactions of the American Philosophical Society* 19, no. 2 (2000): 95–98.

Chiang, John C.H., Yochanan Kushnir, and Alessandra Giannini. "Deconstructing Atlantic Intertropical Convergence Zone Variability: Influence of the Local Cross-Equatorial Sea Surface Temperature Gradient and Remote Forcing from the Eastern Equatorial Pacific." *Journal of Geophysical Research: Atmospheres* 107 (2002): ACL-3-1–ACL-3-19.

Chittick, Neville. "Kilwa and the Arab Settlement of the East African Coast." *The Journal of African History* 4, no. 2 (1963): 179–90.

Colberg, P.M.F., and C.J.C. Reason. "A Model Study of the Angola Benguela Frontal Zone: Sensitivity to Atmospheric Forcing." *Geophysical Letters: Oceans* 33, no. 19 (2006).

Coquery-Vidrovitch, Catherine. *Les routes de l'esclavage: histoire des traites africaines: VIe-XXe siècle*. Paris: Albin Michel, 2018.

Cordeyro, Antonio. *História insulana das ilhas a Portugal sujeitas no oceano ocidental*. 2 vols. Lisbon: Tipografia do Panorama, 1866.

Corriente, Federico. *Dictionary of Arabic and Allied Loanwords: Spanish, Portuguese, Catalan, Galician, and Kindred Dialects*. Leiden, The Netherlands: Brill, 2008.

Coutinho, Gago. "Significação Náutica do Centenário da Descoberta dos Açores." *Conferencia. Seara Nova*, no. 287(1932): 359–63.

Curado, Tiago Gil. "O Ceitil. Um símbolo da Modernidade portuguesa e a sua evolução estética (1440–1570)." *Documenta & Instrumenta* 17 (2019): 19–31.

da Silveira, Ilson C.A., Luiz B. de Miranda, and Wendell S. Brown. "On the Origins of the North Brazil Current." *Journal of Geophysical Research: Oceans* 99, no. C11 (1994): 22501–12.

Davis, William Morris, and William Henry Snyder. *Physical Geography*. Boston: Ginn, 1894.

Day, John. "The Great Bullion Famine of the Fifteenth Century." *Past & Present* 79 (1978): 3–54.

De Lucia, Vito, and Viviana Iavicoli. "From Outer Space to Ocean Depths: The Spacecraft Cemetery and the Protection of the Marine Environment in Areas Beyond National Jurisdiction." *California Western International Law Journal* 49 (2019): 34–90.

DeCorse, Christopher R. "Coastal Ghana in the First and Second Millennia AD: Change in Settlement Patterns, Subsistence and Technology." *Journal des africanistes* 75, *Approches croisées des mondes akan II* (2005): 43–54.

Denamiel, Cléa, W. Paul Budgell, and Ralf Toumi. "The Congo River Plume: Impact of the Forcing on the Far-Field and Near-Field Dynamics." *Journal of Geophysical Research: Oceans* 118, no. 2 (2013).

Dezfuli, Amin. "Climate of Western and Central Equatorial Africa." In *Oxford Research Encyclopedias Africa*. Oxford: Oxford University Press, 2017.

Dias, C. A. "Preliminary Report on the Physical Oceanography off Southern Angola, March and July 1971." *Collection of Scientific Papers International Commission for the Southeast Atlantic Fisheries* 10 (1983): 103–16.

Diop, Salif. *La côte Ouest-Africain du Saloum (Sénégal) à la Mellacoré (Guinée)*. Paris: L'Editions de L'Orstom, 1990.

Domingues, Francisco Contente. *Os navios do mar oceano: teoria e empiria na arquitetura naval portuguesa dos séculos XVI e XVII*. Lisbon: Centro de História da Universidade de Lisboa, 2004.

Doughty, Cheryl L., J. Adam Langley, Wayne S. Walker, et al. "Mangrove Range Expansion Rapidly Increases Coastal Wetland Carbon Storage." *Estuaries and Coasts* 39 (2016): 385–96.

Drouin, Kimberley L., M. Susan Lozier, and William E. Johns. "Variability and Trends of the South Atlantic Subtropical Gyre." *Journal of Geophysical Research: Oceans* 126 (2021): e2020JC016405.

D'Souza, Blanche Rocha. *Harnessing the Trade Winds: The Story of the Centuries-Old Indian Trade with East Africa, Using the Monsoon Winds*. Nairobi: Zand Graphics, 2008.

Düing, Walter, and Friedrich Schott. "Measurements in the Source Region of the Somali Current During the Monsoon Reversal." *Journal of Physical Oceanography* 8, no. 2 (1978): 278–89.

Dynes, Russell R. "The Dialogue Between Voltaire and Rousseau on the Lisbon Earthquake: The Emergence of a Social Science View." *International Journal of Mass Emergencies and Disasters* 18, no. 1 (2000): 97–115.

Ebiatsa-Hopiel-Opiele. "Les teke: définition historique des hommes et de leur espace (avant le XVIIIe siècle)." *Muntu: revue scientifique et culturelle du CICIBA* 7 (1987): 32–47.

Elbl, Ivana. "The Horse in Fifteenth-Century Senegambia." *The International Journal of African Historical Studies* 24, no. 1 (1991): 85–110.

Fairbridge, Rhodes W. "Endospheres and Interzonal Coupling." *Annals of the New York Academy of Sciences* 140, no. 1 (1966): 133–48.

Falchetta, Pero. *Terrarum Orbis: Histoire des représentations de l'espace: textes, images*. Vol. 5 of *Fra Mauro's World Map: With a Commentary and Translations of the Inscriptions*. Begijnhof, Belgium: Brepols, 2006.

Fatoyinboand, Temilola E., and Marc Simard. "Height and Biomass of Mangroves in Africa from ICESat/GLAS and SRTM International." *Journal of Remote Sensing* 34, no. 2 (2013): 668–81.

Fatty, Lamin. "Gambia: Judge in Koina Caste Scuffle Case Threatens to Revoke Accused Persons' Bail Following Their Absence from Court." *Foroyaa* (Gambia), March 22, 2021.

Ferreira, C.M.O. "Os portugueses e a imagem do outro na obra de Álvaro Velho Roteiro da primeira viagem de Vasco da Gama à índia e na Carta a El Rei D. Manuel sobre o achamento do Brasil de Pêro Vaz de Caminha." Master's thesis, Universidade Aberta, 2014.

Ferreira, Susannah Humble, "Juana La Beltraneja, Dynastic Fears, and Threats of Marriage (1475–1506)." *Renaissance and Reformation / Renaissance et Réforme* 43, no. 4 (2020): 79–100.

Finney, Ben R. *Pacific Navigation and Voyaging*. Honolulu, HI: Polynesian Society, 1976.

Finney, Ben R. *Voyage of Rediscovery: A Cultural Odyssey Through Polynesia*. Berkeley: University of California Press, 1994.

Fonseca, Luís Adão da. *O essencial sobre Bartolomeu Dias*. Lisbon: Impr. Nacional-Casa da Moeda, 1987.

Freitas, Isabel Vaz de. *D. Joana, A Excelente Senhora 1462–1530*. Vila do Conde, Portugal: Quidnovi, 2011.

Freitas, Isabel Vaz de. "D. Joana: Uma mulher entre Portugal e Castela." In *En La Europa medieval, mujers com historia, mujers de leyenda Siglos XIII–XVI*, edited by Manuel García Fernández. Seville, Spain: Universidad de Sevilla, 2019.

Garces, Maria Antonia. *Cervantes in Algiers: A Captive's Tale*. Nashville, TN: Vanderbilt University Press, 2002.

Geen, Ruth, Simona Bordoni, David S. Battisti, and Katrina Hui. "Monsoons, ITCZs, and the Concept of the Global Monsoon." *Reviews of Geophysics* 58, no. 4 (2020): 7281–93.

Ghomsi, Franck Eitel Kemgang, Janvier Domra Kana, Zakari Aretouyap, et al. "Main Structural Lineaments of the Southern Cameroon Volcanic Line Derived from Aeromagnetic Data." *Journal of African Earth Sciences* 186 (2002): 104418.

Gill, Adrian E. "Gravity Wave." In *Glossary of Meteorology*, edited by Todd Glickman. Boston: American Meteorological Society, 2000.

Godinho, Vitorino Magalhães. *Os descobrimentos e a economia mundial*. 2nd ed. 3 vols. Lisbon: Editorial Presença, 1981.

Goudie, Andrew S. "Namib Sand Sea: Large Dunes in an Ancient Desert." In *Geomorphological Landscapes of the World*, edited by Piotr Migoń, 163–69. New York: Springer, 2010.

Goudie, Andrew S., and Heather A. Viles. *Geomorphology in the Anthropocene*. Oxford: Oxford University Press, 2017.

Green, Brian, John Marshall, and Aaron Donohoe. "Twentieth-Century Correlations Between Extratropical SST Variability and ITCZ Shifts." *Geophysical Research Letters* 44 (2017): 9039–47.

Green, James. "Material Values of the Teke Peoples of West Central Africa (1880–1920)." PhD dissertation, University of East Anglia, Norwich, 2018.

Hair, Paul H. "The Early Sources on Guinea." *History in Africa* 21 (1994): 87–126.

Hair, Paul H. "The Teixeira da Mota Archive and the Guinea Texts Project." *History in Africa* 10 (1983): 387–94.

Hall, Gwendolyn Midlo. "Africa and Africans in Relational Databases." *The American Historical Review* 115, no. 1 (2010–12): 136–50.

Hall, Kenneth R. *A History of Early Southeast Asia: Maritime Trade and Societal Development, 100–1500*. Lanham, MD: Rowman & Littlefield, 2011.

Hamel, Chouki El. "Race, Slavery and Islam in Maghribi Mediterranean Thought: The Question of the Haratin in Morocco." *The Journal of North African Studies* 7, no. 3 (2002): 29–52.

Higman, Bretwood, Dan H. Shugar, Colin P. Stark, et al. "Is 'Rich-Get-Richer' Valid for the Indian Ocean and Atlantic ITCZ?" *Geophysical Research Letters* 39, no. 13 (2012): L13705.

Huang, Daji, Dingyong Zeng, Xiaobo Ni, Tao Zhang, Jiliang Xuan, Feng Zhou, Jia Li, and Shuangyan He. "Alongshore and Cross-Shore Circulations and Their Response to Winter Monsoon in the Western East China Sea." *Deep Sea Research Part II: Topical Studies in Oceanography* 124 (2016): 6–18.

Ishāk, Halil ibn. *Il lMuntasar "Sommario del diritto malechita."* Translated by Ignacio Guidi and David Santillana. 2 vols. Milan, Italy: Ulrico Hoepli Ditore Libraio della Real Casa, 1919.

Jukes, Thomas. "Historical Perspectives: The Prevention and Conquest of Scurvy, Beri-Beri, and Pellagra." *Preventative Medicine* 18, no. 6 (1989): 877.

Kang, Yan, Delu Pan, Yan Bai, Xianqiang He, Xiaoyan Chen, Chen-Tung Arthur Chen, and Difeng Wang. "Areas of the Global Major River Plumes." *Acta oceanologica sinica* 32 (2013): 79–88.

Kerharo, Joseph, and Jacques G. Adam. "Les plantes medicinales, toxiques et magiques des Niominka et des Socé des lies du Saloum (Senegal)." *Acta Tropica* 25, no. 5 (1968): 279–334.

Khan, Muhammad Akram. *Islamic Economics and Finance: A Glossary*. London: Routledge, 2003.

King, David A. *Islamic Astronomical Instruments*. London: Variorum Reprints, 1987.

Klein, Martin A., and Suzanne Miers. *Slavery and Colonial Rule in Africa*. London: Routledge, 2013.

Klenz, T., M. Dengler, and P. Brandt. "Seasonal Variability of the Mauritania Current and Hydrography at 18°N." *Journal of Geophysical Research: Oceans* 123 (2018): 8122–37.

Kopte, R., P. Brandt, M. Dengler, P.C.M. Tchipalanga, M. Macuéria, and M. Ostrowski. "The Angola Current: Flow and Hydrographic Characteristics as Observed at 11°S." *Journal of Geophysical Research: Oceans* 122, no. 2 (2017): 1177–89.

Koutavas, Athanasios, and Jean Lynch-Stieglitz. "Variability of the Marine ITCZ over the Eastern Pacific During the Past 30,000 Years." In *The Hadley Circulation: Present, Past and Future*, edited by Henry F. Diaz and Raymond S. Bradley, 347–69. Dordrecht, The Netherlands: Springer, 2004.

Kunstmann, Frederick. *Hieronymus Münzer's Bericht über die Entdeckung der Guinea*. Berlin: Verlag der K. Akadamie, 1854.

Kuo, Huei-Ying. "Charting China in the Thirteenth-Century World: The First English Translation of Zhu fan zhi and Its Recipients in China in the 1930s." In *Knowledge in Translation: Global Patterns of Scientific Exchange, 1000–1800 CE*, edited by Patrick Manning and Abigail Owen, 93–116. Pittsburgh, PA: University of Pittsburgh Press, 2018.

Law, Robin. *The Horse in West African History*. Oxford: Oxford University Press, 1980.

Leduc-Leballeur, Marion, Gaëlle de Coëtlogon, and Laurence Eymard. "Air-Sea Interaction in the Gulf of Guinea at Intraseasonal Time-Scales: Wind Bursts and Coastal Precipitation in Boreal Spring." *Quarterly Journal of the Royal Meteorological Society* 139, no. 671 (2013): 387–400.

Lévi-Provençal, Évariste. *Histoire de l'Espagne Musulmane (710–912)*. 3 vols. Cairo: Institut français d'archéologie orientale du Caire, Études historiques, 1945.

Li, C., S. Armstrong, and D. Williams. "Residual Eddies in a Tidal Channel." *Estuaries and Coasts* 29, no. 1 (2006): 147–58.

Lião, Duarte Nunez de. *Origem da Lingoa Portugueza*. Lisbon: Pedro Crasbeeck, 1606.

Lind, James. *A Treatise on the Scurvy in Three Parts*. 2nd ed. London: A. Millar, 1757.

Linhares, Bruno de Andrade. "Aves Marinhas Como Transportadoras de Matéria Para o Ecossistema Insular do Arquipélago dos Abrolho." PhD dissertation, Institute of Oceanography, Universidade Federal do Rio Grande, 2022.

L'Institut de recherche pour le développement. "Histoire de Pointe Noire." *Horizons*. https://horizon.documentation.ird.fr/exl-doc/pleins_textes/divers12–05/13243.pdf.

Luko, C.D., I.T. Simoes-Sousa, J.M. Araujo, and A. Tandon. "Revisiting the Atlantic South Equatorial Current." *Journal of Geophysical Research: Oceans* 126, no. 7 (2021): e2021JC017387.

Mantey, J., F. Owusu-Nimo, K.B. Nyarko, and A. Aubynn. "Operational Dynamics of "Galamsey" Within Eleven Selected Districts of Western Region of Ghana." *Journal of Mining & Environment* 8, no.1 (2017): 11–34.

Marshall, John, Aaron Donohoe, and David Ferreira. "The Ocean's Role in Setting the Mean Position of the Inter-Tropical Convergence Zone." *Climate Dynamics* 42 (2014): 1967–79.

Masiero, Franco. "La raxon de marteloio." *Studi Veneziani* 8 (1984): 393–412.

Massing, Andrea. "Mapping the Malagueta Coast: A History of the Lower Guinea Coast, 1460–1510, through Portuguese Maps and Accounts." *History in Africa* 36 (2009): 331–65.

Matos, Artur Teodoro de, Avelino de Freitas de Meneses, and José Guilherme Reis Leite, eds. *História dos Açores: do descobrimento ao século XX*. Angra do Heroísmo, Portugal: Instituto Açoriano de Cultura, 2008.

Matos, J. Semedo de. "A caravela de Nuño Tristão: A caravela de descobrir." *Revisita de Armada*, no. 218 (1997).

McDonnell, Rachel. "Groundwater Use and Policies in Oman." *International Water Management Group Project Report* 14 (2016).

McDougall, E. Ann. "Harāṭīn in Mauritania." In *Encyclopaedia of Islam*, edited by Kate Fleet, Gudrun Krämer, Denis Matringe, John Nawas, and Devin J. Stewart. 3rd ed. Leiden, The Netherlands: Brill, 2017.

McDougall, E. Ann. "Salts of the Western Sahara: Myths, Mysteries, and Historical Significance." *The International Journal of African History* 23, no. 2 (1990): 231–57.

McIntosh, Susan. "Tools for Understanding Transformation and Continuity in Senegambian Society, 1500–1900." In *West Africa During the Atlantic Slave Trade: Archaeological Perspectives*, edited by Christopher de Corse, 14–37. New York: Bloomsbury Academic, 2001.

McLeod, Elizabeth, Gail L. Chmura, Steven Bouillon, et al. "A Blueprint for Blue Carbon: Toward an Improved Understanding of the Role of Vegetated Coastal Habitats in Sequestering CO2." *Frontiers in Ecology and the Environment* 9, no. 10 (2011): 552–60.

Meeuwis, J. M., and J R. E. Lutjeharms. "Surface Thermal Characteristics of the Angola Benguela Front." *South African Journal of Marine Science* 9, no. 1 (1990): 261–79.

Melzer, B. A., T. G. Jensen, and A. V. Rydbeck. 2019. "Evolution of the Great Whirl Using an Altimetry-Based Eddy Tracking Algorithm." *Geophysical Research Letters* 46, no. 8 (2019): 4378–85.

Mendes, António de Almeida. "Les réseaux de la traite ibérique dans l'Atlantique nord (1440–1640)." *Annales. Histoire, Sciences Sociales* 63 (2008): 739–68.

Mendes-Víctor, Luiz A., Carlos Sousa Oliveira, João Azevedo, and António Ribeiro, eds. *The 1755 Lisbon Earthquake: Revisited*. Dordrecht, The Netherlands: Springer, 2009.

Metcalf, Alida C. *Mapping an Atlantic World, Circa 1500*. Baltimore: Johns Hopkins University Press, 2002.

Michea, H. "De l'Utilisation de Martelloios ou Roses des Vents au Moyen Age." *Navigation*, no. 111 (1989): 101–17.

Miranda, Luiz Bruner de, Fernando Pinheiro Andutta, Björn Kjerfve, and Belmiro Mendes de Castro Filho. *Estuarine Physical Oceanography*. Singapore: Springer Nature, 2017.

Mittelstaedt, Ekkehard. "The Ocean Boundary along the Northwest African Coast: Circulation and Oceanographic Properties at the Sea Surface." *Progress in Oceanography* 26, no. 4 (1991): 307–55.

Mittelstaedt, Ekkehard. "The Upwelling Area off Northwest Africa—a Description of Phenomena Related to Coastal Upwellings." *Progress in Oceanography* 12, no. 3 (1983): 307–31.

Molinari, Robert L., Donald Olson, and Gilles Reverdin. "Surface Current Distributions in the Tropical Indian Ocean Derived from Compilations of Surface Buoy Trajectories." *Journal of Geophysical Research* 95 (1990): 7217–38.

Moreno, Humberto Baquero. "A Exclusão Social no Portugal Medievo." In Maria João Vaz, Eunice Relvas, and Nuno Pinheiro (organizadores), *Exclusão na História. Actas do Colóquio Internacional sobre Exclusão Social*, 9–17. Oeiras, Portugal: Celta Editora, 1999.

Moreno, Humberto Baquero. *Exilados, Marginais e Contestatários na Sociedade Portuguesa Medieval*. Lisbon: Editorial Presença, 1990.

Moreno, Humberto Baquero. *Marginalidade e Conflitos Sociais em Portugal nos séculos XIV e XV*. Lisbon: Editorial Presença, 1985.

Mota, Avelino Texeira da. "A Descoberta da Guiné." *Boletim cultural da Guiné Portuguesa*, no. 1 (1945): 318–21.

Mota, Avelino Teixeira da. *Some Aspects of Portuguese Colonization and Sea Trade in West Africa in the 15th and 16th Centuries*. Bloomington: African Studies Program, Indiana University, 1978.

Munro, John H. "A Maze of Medieval Monetary Metrology: Determining Mint Weights in Flanders, France and England from the Economics of Counterfeiting, 1388–1469." *Journal of European Economic History* 29 (2000): 174–99.

Munro, John H. "Medieval Monetary Problems: Bimetallism and Bullionism." *The Journal of Economic History* 43, no. 1 (1983): 294–98.

Negrão, F., C.H.F. Lacerda, T.H. Melo, et al. "The First Biological Survey of the Royal Charlotte Bank (SW Atlantic). *Estuarine, Coastal and Shelf Science* 255 (2021): 107363.

Nellemann, Christian, Emily Corcoran, Carlos M. Duarte, Luis Valdes, Cassandra De Young, Luciano Fonseca, and Gabriel Grimsditch. *Blue Carbon. A Rapid Response Assessment*. GRID-Arendal, Norway: United Nations Environment Program, 2009.

Neugebauer, Otto. *A History of Ancient Mathematical Astronomy*. Berlin: Springer, 1975.

Newitt, Malyn. *Navigations: The Portuguese Discoveries and the Renaissance*. London: Reaktion Books, 2023.

Okeowo, Alexis. "Freedom Fighter [Biram Dah Abeid]." *The New Yorker*, September 2014

Owusu-Nimo, Frederick, J. Mantey, K.B. Nyarko, E. Appiah-Effah, and A. Aubynn. 2018. "Spatial Distribution Patterns of Illegal Artisanal Small Scale Gold Mining (*Galamsey*) Operations in Ghana: A Focus on the Western Region." *Heliyon* 4, no. 2 (2018): e00534.

Painter, Stuart C. "The Biogeochemistry and Oceanography of the East African Coastal Current." *Progress in Oceanography* 186 (2020): 102374.

Parrain, Camille. "Sailing Routes and Stopovers: Spatial Disparities Across the Atlantic." *Journal of Coastal Research* 61 (2011): 140–49.

Pelegrí, Josep Lluís, and Aïssa Benazzouz. "Coastal Upwelling off North-West Africa." In *Oceanographic and Biological Features in the Canary Current Large Marine Ecosystem*, edited by L. Valdés and I. Déniz-González. Vol. 115 of IOC Technical Series. Paris: IOC-UNESCO, 2015.

Pélissier, Paul. *Les paysans de Sénégal. Les civilisations agraires du Cayor à la Casamance*. Saint-Yrieix-la-Perche: Imprimerie Fabrègue Hutchinson, 1966.

Pereira, Alvaro S. "The Opportunity of a Disaster: The Economic Impact of the 1755 Lisbon Earthquake." *The Journal of Economic History* 69, no. 2 (2009): 466–99.

Peres, Damião. *História dos descobrimentos portugueses*. 4th ed. Porto, Portugal: Vertente, 1992.

Perrone-Moisés, Leyla. *Vinte luas: viagem de Paulmier de Gonneville ao Brasil, 1503–1505*. São Paulo, Brazil: Companhia das Letras, 1992.

Persson, Anders. "How Do We Understand the Coriolis Force?" *Bulletin of the American Meteorological Society* 79, no. 7 (1998): 1373–86.

Peter, K. V., and M. R. Shylaja. "Appendix 1 (ISO List of Plant Species)." In *Handbook of Herbs and Spices, vol. 1*, edited by K. V. Peter. 2nd ed. Oxford: Woodhead, 2012.

Philippon, Nathalie, Guillaume Cornu, Lou Monteil, et al. "The Light-Deficient Climates of Western Central African Evergreen Forests." *Environmental Research Letters* 14, no. 3 (2019).

Picard, Christophe. *La mer et les musulmans d'occident au moyen age VIIe-XIIIe siècles*. Paris: Presses universitaires de France, 1997.

Piper, J. D. A., and A. Richardson. "The Palaeomagnetism of the Gulf of Guinea Volcanic Province, West Africa." *Geophysical Journal International* 29, no. 2 (1972): 147–71.

Pollard, Edward, and Okeny Charles Kinyera. "The Swahili Coast and the Indian Ocean Trade Patterns in the 7th–10th Centuries CE." *Journal of Southern African Studies* 43, no. 5 (2017): 927–47.

Polonsky, Alexander B., Evgeniy G. Nikolaenko, Sekou Konate, and Toumany Camara. "A Study of the Equatorial Atlantic Currents Using Historical Moored Data." *Deutsche Hydrografische Zeitschrift* 45, no. 4 (2017): 219–54.

Poulain, Pierre-Marie. "Coastal Ocean Circulation Experiment off Senegal." U.S. Defense Information Center. 2017. https://www.onr.navy.mil/reports/FY11/popoulai2.pdf.

Ptak, Robert. "From Quanzhou to the Sulu Zone and Beyond: Questions Related to the Early Fourteenth Century." *Journal of Southeast Asian Studies* 29, no. 2 (1998): 269–94.

Pujades i Bataller, Ramon Jose. *Les cartes portolanes: la representació medieval d'una mar solcada*. Barcelona, Spain: Institut cartogràfic de Catalunya, 2007.

Radulet, Carmen. "As Viagens de descobrimento de Diogo Cão: Nova proposta de interpretação." *Maré Liberum* 1 (1990): 181–99.

Raghava Rao, Gobburi Venkatananda. *Scripture of the Heavens: The Eye of the Heavens*. West Bengal, India: Ananda Press, 1949.

Reclus, Élisée. "Peuples des rivières du Sud." In *Nouvelle géographie universelle: la terre et les hommes, vol. 12*. Paris: Hachette, 1876.

Ricard, Robert. “Le commerce de Berbérie et l’organisation économique de l’empire portugais aux XVe et XVIe siècles.” In *Les Annales de l’Institut d’études Orientales de la Faculté des Lettres d’Alger, vol.* 2 (1936): 266–90.

Richardson, Phillip L., and David Walsh. “Mapping Climatological Seasonal Variations of Surface Currents in the Tropical Atlantic Using Ship Drifts.” *Journal of Geophysical Research: Oceans* 91, no. C9 (1986): 10537–50.

Riddle, Joseph Esmond. *The History of the Papacy to the Period of the Reformation.* 2 vols. London: Richard Bentley, 1854.

Risica, Gilda, Alessio Di Roberto, Fabio Speranza, et al. “Reconstruction of the Subaerial Holocene Volcanic Activity Through Paleomagnetic and 14C Dating Methods: El Hierro (Canary Islands).” *Journal of Volcanology and Geothermal Research* 425 (2022): 107526.

Rogers, John H. “Origins of the Ancient Constellations: I. The Mesopotamian Traditions.” *Journal of the British Astronomical Association* 108 (1998): 9–28.

Rogers, John H. “Origins of the Ancient Constellations: II. The Mediterranean Tradition.” *Journal of the British Astronomical Association* 108 (1998): 79–89.

Roller, Duane W. *Through the Pillars of Herakles: Greco-Roman Exploration of the Atlantic*. New York: Routledge, 2006.

Rosell-Fieschi, Miquel, Josep L. Pelegrí, and Jéröme Gourrion. “Zonal Jets in the Equatorial Atlantic Ocean.” *Progress in Oceanography* 130 (2015): 1–18.

Runge, Juergen. “The Congo River, Central Africa.” In *Large Rivers: Geomorphology and Management*, edited by Avijit Gupta, 293–309. Oxford: Wiley Blackwell, 2008.

Sáenz-López Pérez, Sandra. “Dismantling a Myth Through Maps: The Conquest of the ‘Torrid Zone’ from Antiquity to the Renaissance.” In Alfred Hiatt and Carla Lois, “The 23rd International Conference on the History of Cartography Poster Session.” *Imago Mundi* 62 (2009): 98–104.

Santillana, David, and Istituto per l’Oriente. *La comunità musulmana e il suo capo, fonti del diritto e loro ermeneutica, la legge nello spazio e nel tempo, le persone, la famiglia, i diritti reali*. Vol. 2 of *Istituzioni di diritto musulmano malichita: con riguardo anche al sistema sciafiita*. Rome: Anonima Romana Editoriale, 1925–38.

Saraiva, António José. *O crepúsculo da Idade Média em Portugal*. Lisbon: Gradiva, 1990.

Schneider, Ralph R., Peter J. Müller, and Götz Ruhland. “Late Quaternary Surface Circulation in the East Equatorial South Atlantic: Evidence from Alkenone Sea Surface Temperatures.” *Paleoceanography* 10, no. 2 (1995): 197–219.

Schott, Friedrich A., and Jüirgen Fisch. “Winter Monsoon Circulation of the Northern Arabian Sea and Somali Current.” *Journal of Geophysical Research* 105, no. C3 (2000): 6359–76.

Schott, Friedrich A., John C. Swallow, and Michèle Fieux. “The Somali Current at the Equator: Annual Cycle of Currents and Transports in the Upper 1000 m and Connection to Neighbouring Latitudes.” *Deep Sea Research A: Oceanographic Research Papers* 37, no. 12 (1990): 1825–48.

Schottenhammer, Angela. "The 'China Seas' in World History: A General Outline of the Role of Chinese and East Asian Maritime Space from Its Origins to c. 1800." *Journal of Marine and Island Cultures* 1, no. 2 (2012): 63–86.

Schouten, M.W., W.P.M. de Ruijter, P.J. van Leeuwen, and H.A. Dijkstra. "An Oceanic Teleconnection Between the Equatorial and Southern Indian Ocean." *Geophysical Research Letters* 29, no. 16 (2002): 59-1–59-4.

Seed, Patricia. *Ceremonies of Possession in Europe's Conquest of the New World, 1492–1640*. Cambridge: Cambridge University Press, 1995.

Seed, Patricia. "Elefantordenens Portugiske Forbindelse." Translated by Ole Magnus Mølbak Andersen. *Siden Saxo: Magasin for dansk historie* 23, no. 2 (2006): 14–23.

Seed, Patricia. "Navigating the Mid-Atlantic; or, What Gil Eanes Achieved." In *The Atlantic in Global History, 1500–2000*, edited by Jorge Cañizares-Esguerra and Erik R. Seeman, 77–89. Upper Saddle River, NJ: Pearson Prentice Hall, 2007.

Seed, Patricia. *The Oxford Map Companion: One Hundred Sources in World History*. Oxford: Oxford University Press, 2014.

Seed, Patricia. "Rising Sea Levels at Cabo Blanco." In *Esri Map Book 2010*. Redlands, CA: Esri Press, 2010.

Senut, Brigitte, and Loïc Ségalen. "Neogene Palaeoenvironments of the Namib Desert: A Brief Synthesis." *Transactions of the Royal Society of South Africa* 69, no. 3 (2014): 205–11.

Shaffer, Lynda Norene. *Maritime Southeast Asia to 1500*. London: M.E. Sharpe, 1996.

Shahin, Mamdouh. *Water Resources and Hydrometeorology of the Arab Region*. New York: Springer, 2007.

Shetye, S.R., A.D. Gouveia, D. Shankar, S.S.C. Shenoi, P.N. Vinayachandran, D. Sundar, G.S. Michael, and G. Nampoothiri. "Hydrography and Circulation in the Western Bay of Bengal During the Northeast Monsoon." *Journal of Geophysical Research: Oceans* 101, no. C6 (1996): 14011–25.

Shetye, S.R., A.D. Gouveia, S.S.C. Shenoi, G.S. Michael, D. Sundar, A.M. Almeida, and K. Santanam. "The Coastal Current off Western India During the Northeast Monsoon." *Deep Sea Research Part A: Oceanographic Research Papers* 38, no. 12 (1991): 1517–29.

Shonk, Jonathan K.P., Eric Guilyardi, Thomas Toniazzo, Steven J. Woolnough, and Tim Stockdale. "Identifying Causes of Western Pacific ITCZ Drift in ECMWF System 4 Hindcasts." *Climate Dynamics* 50 (2018): 939–54.

Shoshan, Boaz. "Exchange-Rate Policies in Fifteenth-Century Egypt." *Journal of the Economic and Social History of the Orient* 29, no. 1 (1986): 28–51.

Simard, Marc, Lola Fatoyinbo, Charlotte Smetanka, Victor H. Rivera-Monroy, Edward Castañeda-Moya, Nathan Thomas, and Tom Van der Stocken. "Mangrove Canopy Height Globally Related to Precipitation, Temperature and Cyclone Frequency." *Nature Geoscience* 12 (2019): 40–45.

Smith, Andrew B. "Exploitation of Marine Mammals by Prehistoric Cape Herders." *South African Journal of Science* 89 (1993): 161–65.

Smith, Andrew B., and John Kinahan. "The Invisible Whale." *Coastal Archaeology* 16, no. 1 (1984): 89–97.

Soares, Mariza de Carvalho. "'Por conto e peso': o comércio de marfim no Congo e Loango, séculos XV–XVII." *Anais do Museu Paulista* 25, no. 1 (2017): 59–86.

SOS Esclaves. https://www.omct.org/en/network-members/sos-esclaves.

Spiers, Sam. "The Eguafo Kingdom: Investigating Complexity in Southern Ghana." PhD dissertation, Syracuse University, 2007.

Stramma, Lothar, and Friedrich Schott. "The Mean Flow Field of the Tropical Atlantic Ocean." *Deep Sea Research II* 46, nos. 1–2 (1999): 279–303.

Subrahmanyam, Sanjay. *The Career and Legend of Vasco da Gama*. Cambridge: Cambridge University Press, 1997.

Thomas, Nicholas. *Cook: The Extraordinary Voyages of Captain James Cook*. London: Walker, 2003.

Todd, Robert E., Francisco P. Chavez, Sophie Clayton, et al. "Global Perspectives on Observing Ocean Boundary Current Systems." *Frontiers in Marine Science* 6 (2019): 423.

Tomczak, Mattias, and J. Stuart Godfrey. *Regional Oceanography: An Introduction*. Oxford: Pergamon, 1994.

Trenberth, Kevin E., David P. Stepaniak, and Julie M. Caron. "The Global Monsoon as Seen Through the Divergent Atmospheric Circulation." *Journal of Climate* 13 (2000): 3969–93.

Tymowski, Michał. *Europeans and Africans: Mutual Discoveries and First Encounters*. Leiden, The Netherlands: Brill, 2020.

Umukoro, Solomon, and Aderemi Caleb Aladeokin. "Therapeutic Effects of Grains of Paradise (*Aframomum melegueta*) Seeds." In *Nuts and Seeds in Health and Disease Prevention*, edited by Victor R. Preedy, Ronald Ross Watson, and Vinood B. Patel, 535–43. Cambridge: Academic Press, 2011.

University of Rhode Island Graduate School of Oceanography. "Estuarine Physical Oceanography." http://www.po.gso.uri.edu/~codiga/foster/partners.htm.

"The Unspeakable Truth About Slavery in Mauritania." *The Guardian*, June 8, 2018.

U.S. Bureau of Labor Statistics, Commerce Department. "Civilian Occupations with High Fatal Work Injury Rates." 2023.

U.S. Office of the Geographer, Bureau of Intelligence and Research. "Mauritania—Senegal Boundary." *International Boundary Study*, no. 78 (1967).

Valdés, Luis. "Oceanographic and Biological Features in the Canary Current." IOC Technical Series 115, no. 3 (2015): 93–103.

Van Harten, A. M. "Malagueta-peper-een Literatuuronderzoek betreffende de historische Grains of Paradise." *Landbouwkundig Tijdschrift* 177 (1966).

Verlinden, Charles. "Les débuts de la traite portugaise en Afrique (1433–1448)." *Miscellanea mediaevalia in memoriam Jan Frederik Niermeyer*. Groningen, The Netherlands: J. B. Wolters, 1967.

Von Willert, Dieter J., Benno M. Eller, Marinus J. A. Werger, Enno Brinckmann, and Hans-Dieter Ihlenfeldt, eds. *Life Strategies of Succulents in Deserts: With Special Reference to the Namib Desert*. Cambridge: Cambridge University Press, 1992.

Vydrin, Valentin. "Mande Languages." *Oxford Research Encyclopedia of Linguistics*. Oxford: Oxford University Press, 2018.

Vydrin, Valentin. "On the Problem of the Proto-Mande Homeland." *Journal of Language Relationship* 1 (2009): 107–42.

Vydrin, Valentin, Ted G. Bergman, and Matthew Benjamin. "Mandé Language Family of West Africa: Location and Genetic Classification." *Summer Institute of Linguistics Electronic Survey Reports* 2000–003 (2000).

Wang, Bin, Fei Huang, Zhiwei Wu, Jing Yang, Xiouhua Fu, and Kazuyoshi Kikuchi. "Multi-Scale Climate Variability of the South China Sea Monsoon: A Review." *Dynamics of Atmospheres and Oceans* 47, no. 1–3 (2009): 15–37.

Wang, Julie L. McClean, Lynne D. Talley, and Stephen Yeager. "Seasonal Cycle and Annual Reversal of the Somali Current in an Eddy-Resolving Global Ocean Model." *Journal of Geophysical Research: Oceans* 23, no. 9 (2018): 6562–80.

Warren, Bruce A. "Ancient and Medieval Records of the Monsoon Winds and Currents of the Indian Ocean." In *Monsoons*, edited by Jay S. Fein and Pamela L. Stephens, 137–58. New York: John Wiley & Sons, 1987.

Webb, James L. A., Jr. "The Horse and Slave Trade Between the Western Sahara and Senegambia." *The Journal of African History* 34, no. 2 (1993): 221–46.

Wedepohl, P. M., J. R. E. Lutjeharms, and M. Meeuwis. "Surface Drift in the South-East Atlantic Ocean." *South African Journal of Marine Science* 22, no. 1 (2000): 71–79.

Wilcox, Eric M., K. M. Lau, and Kyu-Myong Kim. "A Northward Shift of the North Atlantic Ocean Intertropical Convergence Zone in Response to Summertime Saharan Dust Outbreaks." *Geophysical Research Letters* 37, no. 4 (2010).

Willcox, A. R. *The Rock Art of Africa*. London, Routledge, 2016.

Wright, John. "Slaves, Slavery and the Sahara." In *The Trans-Saharan Slave Trade*, 15–22. London: Routledge, 2007.

Wyffels, Carlos. "Contribution à l'histoire monétaire de Flandre au XIIIe siècle." *Revue belge de Philologie et d'Histoire* 45, no. 4, 1113–41.

Wyffels, Carlos. "Notes sur les marcs monétaires utilisés en Flandre et Artois avant 1300." *Handelingen van het Genootschap voor Geschiedenis* 104, no. 1–2 (1967): 66–87.

Xie, Shang-Ping, and James A. Carton. "Tropical Atlantic Variability: Patterns, Mechanisms, and Impacts." In *Earth Climate: The Ocean-Atmosphere Interaction*, edited by Chunzai Wang, Shang-Ping Xie, and James A. Carton, 121–42. Washington, DC: American Geophysical Union, 2004.

Yemane, Kide, Angesom Gebretsadik, and Yewuhalashet Fissha. "Assessing the Main Challenges of Artisanal and Small-Scale Mining Operations in Africa: A Simple Overview." *Geominetech: The Indian Mineral Industry Journal* 7, no. 2 (2020).

Index

Founded in 1893,
UNIVERSITY OF CALIFORNIA PRESS
publishes bold, progressive books and journals on topics in the arts, humanities, social sciences, and natural sciences—with a focus on social justice issues—that inspire thought and action among readers worldwide.

The UC PRESS FOUNDATION
raises funds to uphold the press's vital role as an independent, nonprofit publisher, and receives philanthropic support from a wide range of individuals and institutions—and from committed readers like you. To learn more, visit ucpress.edu/supportus.